The Woman That Never Evolved

The Woman That Never Evolved

SARAH BLAFFER HRDY

Harvard University Press
Cambridge, Massachusetts, and London, England

10   9   8   7   6   5   4

Library of Congress Cataloging in Publication Data

Hrdy, Sarah Blaffer, 1946–
     The woman that never evolved.

     Includes bibliographical references and index.
          1. Primates—Evolution.   2. Women—Evolution.
3. Feminism.   4. Sociobiology.   5. Sex role.   I. Title.
QL737.P9H79   599.8′0451   81-2921
ISBN 0-674-95540-4 (cloth)   AACR2
ISBN 0-674-95541-2 (paper)

This book is about the female primates who have evolved over the last seventy million years. It is dedicated to the liberated woman who never evolved but who with imagination, intelligence, an open mind, and perseverance many of us may yet become.

Acknowledgments

More than any other person, William Bennett gave me the courage to write this book. At every stage he offered practical advice and provided solutions to problems which arose in the writing. My happiest hours with the manuscript were spent responding to Bill's criticisms. I cannot thank him enough.

I had hardly begun writing when conversations with Robert Bailey, Barbara Smuts, and Richard Wrangham forced me to rethink a number of important issues about female sexuality, cooperation, and competition. I also enjoyed extensive discussions on many topics with Jon Seger. Each of these people gave me valuable insights that worked themselves into the book.

William Bennett, Naomi Bishop, Nancy Burley, Nancy De-Vore, Harry Foster, Daniel Hrdy, James Moore, Jon Seger, Barbara Smuts, Beatrice Whiting, Pat Whitten, and Richard Wrangham each read most or all of the manuscript and offered useful suggestions. Colleagues and close friends showed remarkable tolerance in the face of my defection from the world of science to the demi-world of science writing. Some approved; a few clearly did not. But criticisms from all quarters have improved the final product.

A number of anthropologists, zoologists, and primatologists generously contributed unpublished information and their hard-won special knowledge. These include: Frances Burton, Curt Busse, Tom Butynski, David Chivers, Eric Delson, Mildred Dickemann, Wolfgang Dittus, John Eisenberg, Martin Etter, John Fleagle, Dian Fossey, Annie Gautier-Hion, John Hartung, Paul Harvey, Glenn Hausfater, Clifford Jolly, Eric Keverne, Jane Lancaster, John MacKinnon, Nancy Nicolson, John Oates, Jonathan Pollock, Rick Potts, Rudi Rudran, Joan Silk, Ian Tattersall, Ron Tilson, Erik Trinkaus, Caroline Tutin, Sam Wasser, Beatrice Whiting, John Whiting, Pat Whitten, Kathy Wolf, Pat Wright, and Barbara de Zalduondo. Special thanks are due to Leila Abu-Lughod, Wolfgang Dittus, Joseph Popp, and Richard Wrangham, who provided photographs to use as models for the illustrations.

I thank also Mina Brandt of the Museum of Comparative Zoology library for translation of articles in German, and Nancy Schmidt of the Tozzer Library for help with references. The Peabody Museum provided me with a home base, and I thank especially its director, Karl Lamberg-Karlovsky.

Nancy DeVore of Anthro-Photo helped me locate photographs of rare monkeys. Virginia Savage and Sarah Landry provided the lovely pen-and-ink drawings. Kathleen Horton typed several drafts of the manuscript. I am grateful to these women for much more than just the tangible results of their labor. Our friendships date back nearly a decade, and it has been a pleasure to once again enlist their help.

I thank my teachers, particularly Irven DeVore, Robert Trivers, and Edward O. Wilson, who drew me into the field of evolutionary biology and who taught me evolutionary theory. The years between 1971 and 1975 were heady times for those of us then graduate students in the life sciences at Harvard. The discipline of sociobiology, conceived earlier by R. A. Fisher, W. D. Hamilton, and George Williams, was approaching term. According to a book review by Mary Jane West Eberhard, the field was "born" in 1975 with the publication of Wilson's book *Sociobiology: The New Synthesis,* and, as is widely known, that child attracted a great deal of passionate attention. Yet it would be misleading to remember only the intellectual excitement of those years. Within the Harvard of that time there was no overlap at all between feminism and evolutionary biology, not even a common language. Feminists were outraged at what they took the sociobiologists to be saying, and the sociobiologists were mystified to discover that feminists were demonstrating at their lectures. As a woman in the midst of all this, I felt torn and often quite alone. I owe a special debt to the writings of Antoinette Brown Blackwell, Jean Baker Miller, Niles Newton, Katherine Ralls, and Mary Jane Sherfey, which were truly beacons in the night to me. In their different ways, they pioneered a feminist perspective on biology, a perspective that was simply not available to me from my formal education.

Then there are acknowledgments of another sort. My husband, Daniel Hrdy, was, as always, steadfast in his support. He alone knows the extent of my debt to that peculiarly human institution of the companionate marriage. During the four years that it took to research and write this book, our daughter was often in the care of people other than her parents. I extend heartfelt thanks to Marjorie Delaney, Diane Lusk, Myra Bennett, and other parents and staff who work at her daycare center. My own hours working in that communal environment provided therapeutic balance to writing about females in a harsher world. Such centers are of course gardens of privilege for a privileged species in a privileged portion of the globe where many of the ordinary rules of natural selection have temporarily been suspended. For the time being my daughter is part of a world where there is plenty for everyone and where she enjoys the luxury of a social environment in which equal opportunity for each individual is respected.

S.B.H.

Contents

The Woman That Never Evolved

Natural selection is not always good, and depends (see Darwin) on many caprices of very foolish animals. GEORGE ELIOT, 1867

1

Some Women That Never Evolved

Biology, it is sometimes thought, has worked against women. Assumptions about the biological nature of men and women have frequently been used to justify submissive and inferior female roles and a double standard in sexual morality. It has been assumed that men are by nature better equipped to conduct the affairs of civilization, women to perpetuate the species; that men are the rational, active members of society, women merely passive, fecund, and nurturing. Hence, many readers will open a book about the biology of female primates with considerable apprehension.

Feminists in particular may rebel at the thought of looking to the science of biology for information that bears on the human condition. They may be put off by the fact that among our nearest relations, the other primates, the balance of power favors males in most species. Yet, if they persist, readers may be surprised by what else they learn concerning their distant cousins and, by inference, their own remote ancestresses. They will find no basis for thinking that women—or their evolutionary predecessors—have ever been dominant over men in the conventional sense of that word, but they will find substantial grounds for questioning stereotypes which depict women as naturally less assertive, less intelligent, less competitive, or less political than men are.

For at least two reasons, feminists have tended to reject biological evidence about females of other species in their thinking about the human condition. First, there is a widespread misconception that "biology is destiny."[1] According to this view, if even a portion of the human male's dominance is ascribed to evolutionary causes, an intolerable status quo will have to be condoned as fundamentally unalterable. Second, biological evidence has been repeatedly misused to support ideological biases, and field studies have been designed and executed in the thrall of such biases. Certainly, this has been the case in the study of other primates. Research has focused on the way adult males maneuver for dominance while females attend to the tasks of mothering; it has neglected the manifestations of dominance and assertiveness in females themselves, behavior that sometimes brings females into conflict with males and with each other.

Primatology is a rapidly expanding field. The most accurate information about female primates has only been collected in the last decade. Much of it is confined to Ph.D. theses and technical accounts and has yet to find its way into the mainstream of the social sciences. Disastrously, experts writing about sex differences among primates have relied upon stereotypes of the female primate constructed in the early sixties.[2] Pretend this is a quiz. Which of the following recent statements about primate social structure, all made by eminent social scientists, also happen to be obsolete?

> "The dominant male is obviously the central figure in the group's persistence over time."[3]
> "Competition is peculiar to the male sex."[4]
> "There is reason to believe that the female hierarchies are less stable. A female's status tends to change when she is in estrus, and to reflect the status of her male consort while she is in the mating phase of her cycle."[5]

The answer is that all three are out of date. Yet such stereotypes have led to the widespread impression that "primate females seem biologically unprogrammed to dominate political systems, and the whole weight of the relevant primates' breeding history militates against female participation in what we can call 'primate public life.' "[6] As we shall see in the course of this book, few statements about the biological origins of sexual asymmetries could be quite so far from the truth.

An occasional voice has warned that there was another side to this story—the work of the anthropologist Jane Lancaster comes to mind[7]—but the reports about competitive males and mothering females continue to roll out of the textbook mills and are currently entrenched in college curricula and in popular literature. By comparison, more accurate accounts are technical and less accessible. Not surprisingly, otherwise broadminded writers and policymakers in psychology and the humanities (particularly those sympathetic to feminist goals) have ignored the primate record or chosen to reject it altogether.[8] As a curious result, today we find that theories explaining the nearly universal dominance of males fall into two categories: hypotheses that are either biologically oriented and informed by stereotypes (that is, views which contain a kernel of truth but are, on the whole, quite misleading), or those that eschew the primate evidence altogether and thereby ignore much that is relevant to understanding the human condition.

When I refer to dominance among humans, I mean the ability to coerce the behavior of others. Among nonhuman primates, a simpler definition is often feasible because dominance hierarchies can be recognized from observations of one-on-one interactions between individuals competing for the same desired resource. When speaking of nonhuman primates, then, I use "dominant" to describe the animal that usually wins in a one-on-one encounter, the animal that typically can approach, threaten, and displace another. No one is particularly satisfied with the concept of dominance. Typically, dominance is difficult to assess and highly dependent on context; furthermore, dominance is not necessarily related among different spheres of activity. Hence, the publicly acclaimed emperor may be ruled by his wife at home; a sated tyrant may lose a wedge of meat when matched against a particularly hungry minion; and the richest or most powerful male may not beget the most children if his wives are routinely unfaithful. Nevertheless, the ability of one individual to influence or coerce the behavior of others, usually by threatening to inflict damage but also by promising to give (or withhold) rewards, remains a real phenomenon, and a term for it is useful. Even the most ardent critics of the concept do not advocate total expurgation of the term.

Whatever definition we might choose, though, there seems to be a general consensus among anthropologists that for most human societies, sexual asymmetry appears in dominance relations, and it gives the edge to males. Hence,

> Whereas some anthropologists argue that there are, or have been, truly egalitarian societies . . . and all agree that there are societies in which women have achieved considerable social recognition and power, none has observed a society in which women have publicly recognized power and authority surpassing that of men . . . Everywhere we find that women are excluded from certain crucial economic or political activities . . . It seems fair to say, then, that all contemporary societies are to some extent male-dominated, and that although the degree and expression of female subordination vary greatly, sexual asymmetry is presently a universal fact of human social life.[9]

The obvious question is, Why?

Psychologists and anthropologists have proposed a variety of explanations for male domination among humans. The following is not an exhaustive list, but it includes the major current theories.

Following Marx and Engels, one scenario begins with an egalitarian species. Only when an economic transition facilitated the accumulation of surpluses and trade, which in turn led to warfare in the defense of material goods and trade routes, did women lose out. As valuable producers but inferior warriors, they yield their autonomy to male capitalists.[10]

Post-Freudian theory holds that subordination of women results from conditions of socialization. Long periods of close association between mother and offspring foster close identification of daughters with their mothers. Whereas daughters fail to form any strong sense of separate identity, boys must struggle to define their own gender role, and in the process not only deny but also devalue all that seems feminine.[11]

Anthropologists from the structural school tell us that people associate women and their procreative functions such as menstruation and childbirth with Nature and natural processes. By contrast, men are identified with Culture and civilized processes. Because people perceive Culture to be superior to Nature, females by analogy are perceived as inferior.[12]

For many "biobehaviorists," it was Man the Hunter who usurped the independence of women: big-game hunting, a peculiarly human adaptation, led to social inequality between the sexes. In one widely cited version of the theory, as hunting became important, the strength of males combined with their freedom from encumbering babies quickly permitted them to monopolize the chase and the distribution of meat. Success depended on special male skills: visual–spatial capacities, stamina, stalking abilities, and especially cooperation. According to a now notorious extension of this scenario, "our intellect, interests, emotions, and basic social life—all are evolutionary products of the success of the hunting adaptation."[13] (Curiously, few anthropologists have asked why intelligence never became sex-linked or why—if intelligence evolved among males to help them hunt—Nature should have squandered it on a sex that never hunted.) The hunting hypothesis was later refined to emphasize the importance of male predispositions to bond with other males: such bonds provided the power base for subsequent political preeminence achieved by men.[14] Furthermore, male hunters were able to cement reciprocal relations with an even wider network of allies through the presentation of meat. Men engendered obligations and gained recognition by such gifts. Once male preeminence was established, females themselves became objects of exchange and were given in marriage by brothers or fathers who received wives for themselves in return.[15]

Although essentially male-centered and to some tastes "sexist," these theories rely on traditional anthropological tenets. Feminist reconstructions of this stage in human evolution are based on the same assumptions about early human ecological adaptations; they also focus on division of labor, sharing, the right to allocate resources, and the importance of ritual bonding. (Theory has even found its way into practice. Feminist educators, for example, have absorbed the notion that in order to compete successfully for power, women's socialization must begin to incorporate the lessons and social reflexes to be learned from teamwork. In a recent book on managerial women, the authors advocate competitive team sports so that women leaders-in-training may participate in this contemporary analogue of hunting and tribal warfare.) [16] In developing a new perspective, revisionists highlight female contributions to subsistence, tool manufacture, and cultural traditions, but they leave the basic outlines of early human ecology unchanged. For example, the feminist anthropologists Adrienne Zihlman and Nancy Tanner concur with the conventional view that people diverged from other primates around five million years ago, and they hypothesize that as early humans shifted from forest to savanna they increasingly shared resources, differentiated assignment of task by sex, and relied on tools. Zihlman and Tanner regard these changes as central to the transformation of our primate ancestors, but they also emphasize that women were gathering a large proportion of the food, that the vegetables women gathered were crucial to subsistence, and that it was women who tended to invent new food-getting technologies and to transmit this information from generation to generation. [17] It is different wine in the same bottle: now woman is the toolmaker. From this perspective, male "superiority" is simply an impression conveyed by biases in data collection and analysis. [18]

Here, then, are five theories to explain male dominance, each highly informative in its own right. But they all share one striking deficiency. Each focuses upon the human condition and lays the burden of sexual inequality, real or mythical, at the doorstep of specifically human attributes: the production of surpluses and the subsequent rise of trade economies; the discovery of the "self" and the formation of ego boundaries; binary conceptualizations of the universe which engender oppositions such as Nature and Culture; big-game hunting; and a sexual division of labor related to subsistence. Each of these theories may contribute to our understanding of the human case, but even taken together, they are insufficient to explain the widespread occurrence of sexual inequality in nature, inasmuch as they account for only a small portion of known cases. They cannot explain sexual asymmetry in even one other species. Yet male dominance characterizes the majority of several hundred other species that, like our own, belong to the order Primates. Save for a handful of highly informative exceptions, sexual asymmetries are nearly universal among primates. Logic alone should warn us against explaining such a widespread phenomenon with reference only to a specialized subset of human examples.

IT IS of course completely appropriate in some respects that theories to explain the peculiar status of women relative to men should focus as they do on uniquely human attributes. We do differ from other animals in our use of language, in our creation and transmission of value systems and advanced technologies, and, most importantly, in our capacity to formulate and articulate conscious decisions. Other creatures simply fall into place within social systems that persist because they happen to be evolutionarily stable. We, by contrast, exhibit an insatiable desire to imagine or bring about novel social systems, some of them idealistic or even utopian in character.

So our idealism—and our ability to consciously change our society—sets us apart from other creatures, but that does not give us license to devalue the facts about other primates. Indeed, awareness of the differences, when combined with knowledge of our close relation by common descent with the other apes, ought instead to make us wonder out loud how we could have come to be the way we are. Although opinions differ as to whether chimpanzees or gorillas are our closest living relatives, it is clear that we are more closely related to these two species of great apes than either chimps or gorillas are to the third great ape, the orangutan. By current estimates, only five million years have elapsed since the nearest common ancestor we share with chimps. The genes of humans and chimps are biochemically almost indistinguishable—a fact which has led scientists to suspect that a relatively small number of genes governing the timing of development make all the difference between speaking, culture-bearing humans and our less talkative cousins.[19]

There is an impressive degree of continuity in the experience of humans and other higher primates (this includes both monkeys and apes) that goes far beyond similar anatomy and biochemistry, fingernails, and stereoscopic vision. We and the other higher primates perceive the world in a similar fashion, and we process information in similar ways. For example, we share striking neuroanatomical patterns in those portions of the brain concerned with memory.[20] Expressions of emotion, such as the smile, can be traced from species to species and identified in very rudimentary form in the "open-mouth display" of other primates.[21] Under appropriate conditions female primates, from hamadryas baboons living in a harem to women living in college dormitories, tend to synchronize their menstrual cycles. At the beginning of the school term, young women arriving from all parts of the country are cycling on different schedules; by the end of the school year, close friends menstruate around the same time of the month.[22] Several recent studies have shown that women, like other primates, are more likely to initiate sexual activity around the time of ovulation (a controversial finding, discussed in detail in Chapter 7), and there is increasing evidence that other aspects of woman's sexuality, such as her capacity to experience orgasms (Chapter 8), are shared by other primate females. Most importantly, as we shall learn in Chapters 5 and 6, it is competition among individuals of the same sex (not just competition among males, but also among females) that has permitted reproductive exploitation of one sex by the other to evolve and be maintained (that is, a member of one sex manipulating another to his or her own reproductive advantage). In this respect, humans may be far more similar to other primates than we are different from them.

On the other hand, by refusing to talk about biology, we effectively hide the fact that there are important ways in which human females are in a worse position than are females in other species. (One of the justifications, after all, for ignoring the animal evidence is that supposedly it paints a picture prejudicial to the aspirations of women.) Among humans there is a universal reliance on shared or bartered food. In many societies, a woman without a man to hunt or earn income, or a man without a wife to do the cooking, is at considerable disadvantage. By contrast, among all nonhuman primates each adult is entirely responsible for supplying his or her own food. The only exceptions involve occasional meat sharing among chimpanzees, but even here males tend to monopolize meat from cooperatively hunted prey (small ungulates and other primates); females rely for animal protein on termites and other individually obtained small prey. Among chimpanzees there is a rudimentary division of labor by sex, but in no case does one sex depend on the other for any staple.[23]

In this respect, female primates (and also, one could argue, the males) enjoy greater autonomy than do either men or women. In roughly 80 percent of human societies, fathers or brothers exercise some control over adolescent and adult females. Such *authority* does not exist among other primates. The Marxists have a point: patriarchy tends to develop where women produce commodities and not just offspring. We have a uniquely elaborate division of labor by sex, and a unique reliance on sharing.[24] But more basic asymmetries between the sexes, based on reproductive exploitation of one sex by the other, long predate the human condition. The fact that males are almost universally dominant over females throughout the primate order does not mean that males escape being used! But dominant they are, with only a few (very important) exceptions (Chapters 3 and 4). Since we are typically primate in this respect, it seems foolish to continue to focus our attention exclusively on those features of our way of life in which we are *untypical* of other primates. Male authority is indeed uniquely human, but its origins are not.

PRIMATOLOGISTS tend to see the world a bit differently from other people. Not surprisingly—it's an odd occupation, after all, crawling under brambles to keep a monkey or an ape in view. Primatologists pay attention to what animals do, not to what they say they do. And primatologists tend to be excessively curious about ancestry. What sort of ancestor did the creature at hand evolve from? And why? What social and environmental pressures made it advantageous for an individual to possess a certain trait? Because of the taxonomic

relationship between us and the other primates, few primatologists can resist the temptation to combine an anthropocen-
tric concern for *Homo sapiens* with this urge to understand
origins. A peculiar perspective, no doubt, but it is my contention that a broader understanding of other primates is going
to help us to expand the concept of human nature to include
both sexes, and that it is going to help us to understand the
problems we face in attempting to eliminate social inequalities based on sex. In the process, we will also find out why
some current notions of what it means to be female depict na-
tures that never did, and never could have, evolved within the
primate lineage.

For example, the belief that women once ruled human affairs still enjoys a certain currency among some feminists,
particularly those who work in a Marxist tradition. They inherit the notion, by way of Friedrich Engels, from a Swiss
jurist and student of Roman law, Johann Bachofen. Supporting his ideas with copious references to ancient mythology
embellished with bits of archaeology and pre-Hellenic history,
Bachofen in 1861 published an outline of human history entitled *The Law of the Mother (Das Mutterrecht)*. In it, he proposed that people first lived in a state of cheerful promiscuity
which then gave way to a more orderly society controlled by
women. Matriarchy was supplanted, gradually, by systems in
which men were dominant, and those have persisted until the
present. Bachofen believed that a matriarchal phase was universal in the history of human societies and was not a special
adaptation to environmental or political circumstances.[25]

Yet the weight of evidence from anthropology and archaeology since Bachofen's time has not favored his view. To be
sure, there have been societies in which property was passed
through the female line and children were identified primarily
as their mother's offspring rather than their father's. Such
matri*lineal* (not matri*archal*) arrangements are far from rare
among human societies. About 15 percent of the world's cultures reckon inheritance through mothers, and in about half
of these a man goes to live with his wife's family when he
marries. (As a rule these societies are horticultural, and the
property in question is a garden plot passed from

mother to daughter.) Yet even in these circumstances men tend to become the administrators of the family's wealth and retain the governing voice in collective affairs.[26] It is certainly possible that some groups of women banded together to live like Amazons, but such societies were never a universal stage in human evolution.

Myths about women ruling the world usually come linked with a theory about the true nature of women. The prototypical matriarchs, the Amazons, were believed to be on the whole aggressive and warlike—masculine spirits in drag. At the other extreme, the idealized women of *Herland*—Charlotte Perkins Gilman's marvelous 1915 utopian novel about an all-female society—were even-tempered and utterly rational creatures whose solidarity dumbfounded a male spy into exclaiming, "Women can't cooperate—it's against nature."[27] Both traditions have recent exponents. Valerie Solanis revived the Amazonian ethos in her 1967 manifesto for the Society for Cutting Up Men (SCUM),[28] while Elizabeth Gould Davis refurbished Gilman's vision in her book *The First Sex*, which averred that there once was a "golden age of queendoms, when peace and justice prevailed on earth and the gods of war had not been born."[29]

The matriarchal fallacy and the myths linked with it about the nature of women are not merely a misreading of the anthropological and paleontological records. They have also provided a refuge from and a defense against another, more popular nineteenth-century belief about the nature of women: that they are sexually passive creatures devoted to the tasks of mothering and that they are devoid of political instincts. This doctrine of female inferiority has disfigured several ostensibly impartial realms, particularly the study of human evolution. Such ideas have predisposed biologists to some curious conclusions about women and female animals in general. For example, it is often assumed—most often implicitly—that only males gain an evolutionary advantage from being competitive or sexually adventurous. To the extent that female behavior contradicts these assumptions (the subject of Chapter 7), it is dismissed as merely a by-product of the masculine character.

I am scarcely the first person to point out that the evolution of female traits is no less subject to the rigors of competition than that of males. Just four years after Charles Darwin published *The Descent of Man and Selection in Relation to Sex* (1871), Antoinette Brown Blackwell published a polite critique of the book. She made no bones about her commitment to both feminism and to Darwin's theories about natural selection. But Blackwell wished that she could broaden his perspective.

> Mr. Darwin, also, eminently a student of organic structures, and of the causes which have produced them, with their past and present characters, has failed to hold definitely before his mind the principle that the difference of sex, whatever it may consist in, must itself be subject to *natural selection* and to evolution. Nothing but the exacting task before him of settling the Origin of all Species and the Descent of Man, through all the ages, could have prevented his recognition of ever-widening organic differences evolved in two distinct lines. With great wealth of detail, he has illustrated his theory of how the male has probably acquired additional masculine characters; but he seems never to have thought of looking to see whether or not the females had developed equivalent feminine characters.[30]

In accepting the theory of natural selection, Blackwell firmly rejected the doctrine of female inferiority and the idea that females are somehow incomplete versions of males—beliefs which "need not be accepted without question, even by their own school of evolutionists." But the evolutionists were not listening.

In the late nineteenth century the popular understanding of evolution became permeated by social Darwinism, a philosophy most closely identified with Herbert Spencer, who was energetically adapting Darwin's theories to fit his own political views. Spencer thought females never had been inherently equal to males and could never be; subordination of women was not only natural but, in his view, desirable.[31]

Social Darwinism has, almost indelibly, tainted most people's understanding of evolutionary theory—certainly as it applies to human beings. Yet social Darwinism differs from Darwinism-without-adjectives in one all-important way, and ignoring this distinction has been one of the most unfortunate and long-lived mistakes of science journalism. Darwinism proper is devoted to analyzing all the diverse forms of life according to the theory of natural selection. Darwinists describe competition between unequal individuals, but they place no value judgment on either the competition or its outcome. Natural-selection theory provides a powerful way to understand the subordination of one individual, or group of individuals, by another, but it in no way attempts to condone (or condemn) subordination.

By contrast, social Darwinists attempt to *justify* social inequality. Social Darwinism explicitly assumes that competition leads to "improvement" of a species; the mechanism of improvement is the unequal survival of individuals and their offspring. Applying this theory to the human condition, social Darwinists hold that those individuals who win the competition, who survive and thrive, must necessarily be the "best." Social inequalities between the sexes, or between classes or races, represent the operation of natural selection and therefore should not be tampered with, since such tampering would impede the progress of the species. It is this latter brand of Darwinism that became popularly associated with evolutionary biology. The association is incorrect, but it helps to explain why feminists have steadfastly resisted biological perspectives.

Blackwell's informed dissent was drowned out in the wake of popular acceptance of social Darwinism. Her contribution to evolutionary biology can be summed up with one phrase: the road not taken. This turning point, over a century ago, left a rift between feminism and evolutionary biology still not mended. Historically and politically, there was obvious justification for the split. There has been a prevailing bias among evolutionary theorists in favor of stressing sexual competition among males for access to females at the expense of careful scrutiny of what females in their own right were doing. Among their recurring themes are the male's struggle for preeminence and his quest for "sexual variety" in order to inseminate as many females as possible. Visionaries of male–male competition stressed the imagery of primate females herded by tyrannical male consorts: sexually cautious females coyly safeguarding their fertility until the appropriate male partner arrives; women waiting at campsites for their men to return; and, particularly, females so preoccupied with motherhood that they have little respite to influence their species' social organization. Alternative possibilities were neglected: that selection favored females who were assertive, sexually active, or highly competitive, who adroitly manipulated male consorts, or who were as strongly motivated to gain high social status as they were to hold and carry babies. As a result, until just recently descriptions of other primate species have told little about females except in their capacity as mothers. Natural histories of monkeys and apes have described the behavior of males with far greater detail and accuracy than they have described the lives of females. Small wonder, then, that audiences sensitized to both the excesses of social Darwinism and conventional sexism have found this emphasis upsetting.

Yet evolutionary biology, and its offspring, sociobiology,[32] are not inherently sexist. The proportion of "sexists" among their proponents is probably no greater than the proportion among scientists generally. To be sure, contemporary analyses of mammalian breeding systems can cause even a committed Darwinian like myself to contemplate her gender with foreboding. Yet, it is all too easy to forget, while quaking, that sociobiology, if read as a prescription for life rather than a description of the way some creatures behave, makes it seem bad luck to be born either sex.

T HE purpose of this book, then, is to dispel some long-held myths about the nature of females, and to suggest a few plausible hypotheses about the evolution of woman that are more in line with current data. Throughout the discussion, it will be well to keep in mind a central paradox of the human condition—that our species possesses the capacity to carry sexual inequality to its greatest known extremes, but we also possess the potential to realize an unusual social equality between the sexes should we choose to exercise that potential. However, if social inequality based on sex is a serious problem, and if we really intend to do something constructive about it, we are going to need a comprehensive understanding of its causes. I am convinced that we will never adequately understand the present causes of sexual asymmetry in our own species until we understand its evolutionary history in the lines from which we descend. Since we cannot travel back in time to see that history in the making, we must turn to those surrogates we have, other living primates, and study them comparatively. Without the perspective such a study affords, we will remain ignorant of the most fundamental aspects of our own situation, in part because of a diminished ability to ask interesting questions about it.

No comradely socialist legislation on women's behalf could accomplish a millionth of what a bit more muscle tissue, gratuitously offered by nature, might do."   ELIZABETH HARDWICK, 1966

2

An Initial Inequality

It has long been customary among members of our species to assume that males are dominant over females. Male dominance need not be regarded as inevitable in order to admit that it has been a potent social concept. Among other species in the primate order (which includes apes, monkeys, lemurs, and a few furry odds and ends), females have far more power than has traditionally been acknowledged. Nevertheless, only exceptionally do females exercise direct control over male behavior. By contrast, in generation after generation, species after species, or in the human case, culture after culture, primate males have been able to dominate females and to translate superior fighting ability into political preeminence over the seemingly weaker and less competitive sex. Why?

This question was not foremost in my mind when I switched from the study of people to the study of monkeys, but neither was it inaccessibly buried in my subconscious. It was not an accident that I was drawn to the study of the Hanuman langur, a species characterized by ruthless exploitation of females by males. Rather, I suspect that I was drawn to this depressing topic by the same desire that makes victims in a disaster who know that the situation is bad want to hear the very worst. I have sometimes compared my nine years of observing and writing about these sleek silver monkeys of India to watching a decade-long performance of a play by Strindberg, a writer who believed that males and females were as dissimilar as "two different species," and who some say went mad from dwelling too long upon irreconcilable differences between the sexes.

In the course of my expedition among the langurs I learned that male exploitation of females is not always as clear-cut as it seems, as I hope to show in Chapter 7. Though years of studying langurs prepared me for much of what I was to learn about male–female relations in other species as I started to research this book, I was not prepared for the degree to which the plight of langurs, in one form or another, was so nearly universal among primates. Why this omnipresent inequality?

Among langurs, it is females, not males, who comprise the stable core of social organization; as with most monkeys, langur society is given its shape by the relationships between overlapping generations of related females. In all but a few species, females are permanent residents in social groups, males mere transients. In fact, the first evolutionary step toward social life for any mammal is thought to have taken place when related females began to develop mutual tolerance and to cooperate with each other. In those primates where rank is inherited from the mother, female dominance relations have far more long-term influence than does the ephemeral power politics of males, and the rank an individual's mother happens to hold may be the single most important fact of its biological existence.

Most monkeys like langurs or macaques live on real estate inherited by a daughter from her mother. When colonization of new habitats occurs, the role of Moses may be taken by a low-ranking older female who leads her disadvantaged lineage into the wilderness. Most such splinters perish, but those that survive and breed extend the boundaries of the species. From one such female foundress there may one day originate a new species; Moses becomes Eve.

Both in new habitats and on home territory, females—often old ones—provide the reservoir of knowledge and remembered traditions needed if the troop is to survive from season to season in the face of environmental vagaries. The oldest veteran of any monkey troop is likely to be a female, and it is she who will remember the distant waterhole in a drought. To the extent that such traditions can be considered culture (or protoculture), females are its chief transmitters. Knowledge is passed by a mother to her offspring.

Half of the genes of any population derive from females. As with all mammals, the breeding success of each male primate depends on extensive and long-term investment in offspring by females. Moreover, to an extraordinary degree, the predilections of the investing sex—females—potentially determines the direction in which the species will evolve. For it is the female who is ultimate arbiter of *when* she mates and how often and with whom. This is especially true in the case of primates. Rape is unknown except in people and in one of the great apes, the orangutan. Females, then, are and always have been the chief custodians of the breeding potential of the species. All the more puzzling, then, is the seemingly ubiquitous inequality between the sexes.

Despite the obvious importance of females to the well-being and evolution of the species, with few exceptions males have the advantage over females in any direct conflict. To be sure, confrontations between males and females are often prevented by social arrangements, the two sexes existing, to a great extent, in different dimensions of the same world. Hierarchies, for example, are often separate, males competing with males, females with other females. Nevertheless, on those occasions when a male and a female covet the same fig or the same safe crotch of a tree to spend the night, it will typically be the male who gets it.

In the majority of primates, males are bigger than females and are able to bully them. Exceptions to this pattern are few, yet more are known today than any professional primatologist dreamed possible even a decade ago. Where special circumstances decrease the advantage of aggressiveness among males, females are able—for all or part of the year—to dominate them. And in the case of species whose males participate almost equally with females in rearing offspring, the female may be equal in size or even bigger. The scarcity of such iconoclastic species belies their significance.

Clearly the potential for doing things differently exists. The evolution of bigger females, or dominant females, does not defy gravity or any other natural law. In most invertebrates and in many fish, females are larger than males. The phenomenon is less common among mammals, but it does occur. The female of the lesser wrinkled-face bat is so much bigger than her male counterpart that for years the two sexes were classified as different species. Indeed, the biggest mammal (or any other creature) that ever lived was probably female, since females are larger than males in the record-holding blue whale. We owe our awareness of mammals in which females are bigger than males to the zoologist Katherine Ralls, who painstakingly unearthed from a scattered literature all the cases she could locate.[1]

Rall's work makes a very important point because it undermines the view that larger size alone automatically leads to dominance. Sociologists and zoologists have usually taken the position that males dominate females simply because they are bigger than females, and indeed there is nothing parochial about this hypothesis; it applies to a broad spectrum of animals and deserves to be examined closely. As the sociobiologist Donald Symons sums up this widely shared viewpoint:

> Chimpanzee male dominance is clearly a function of size, strength, and aggressiveness; the same probably is true of humans, but one can ignore the issue of aggressiveness and still explain human male dominance . . . the importance of physical strength in human affairs should not be underestimated . . . To explain male political dominance, all one needs to assume is that individuals tend to use the most effective tools they have at their disposal to get what they want, and that males choose other males rather than females as political allies because males make more effective allies.[2]

The physical factor in male dominance is dismissed out of hand by some feminists, but rarely are they able to present convincing evidence for doing so.[3] By contrast, Ralls provides unusually hard and persuasive reasons for treading cau-

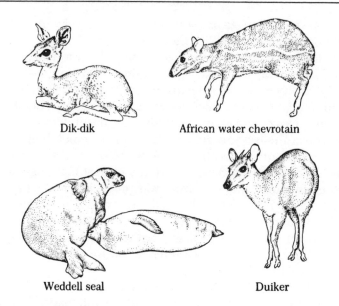

Dik-dik   African water chevrotain

Weddell seal   Duiker

tiously in this domain. Although she does not rule out body size as a factor, she cites a variety of cases (the dik-dik, the duiker, the golden hamster, African water chevrotain, and Weddell seal) to prove that even among those mammals in which the *females* are bigger, males remain the more aggressive—and presumably dominant—sex. Furthermore, among primates of both sexes, many examples can be found of a smaller individual able to dominate a larger one of the same sex,[4] and certainly similar cases are known for humans. It is premature, and quite possibly wrong, to dismiss body size as a factor in male dominance; however, it is clearly not the only, and may not even be the major, factor at work.

The rare primate species in which females are bigger than males can be counted on one hand. All belong to the family Callitrichidae, composed of marmosets and tamarins. Species such as ring-tailed lemurs, sifakas, or talapoin monkeys in which females are equal in size to males or smaller but nevertheless dominate them are scarcely more numerous. And here again, the question is why? What barriers in the biological and social arrangements of primates limit the proliferation of societies with overtly powerful females?

Part of the answer, remote in time and nonnegotiable—but not particularly satisfying—is *anisogamy*, from the Greek *aniso* (meaning unequal) plus *gametes* (eggs and sperm): gametes differing in size.[5] Sex, it is now thought, began as a simple act of hijacking when, some several billion years ago, a small cell waylaid and merged with a bigger one, richer in substance and nutrients. What began as a chance, quantitative difference in bulk became eventually a difference in strategy and kind. Competition among small cells for access to the largest ones favored smaller, faster, and more maneuverable cells, analogous to sperm. The hostages we might as well call ova. The egg provided not only half of the genetic material but the resources to sustain development. The ground rules for the evolution of two very different creatures—males and females—were laid down at this early date. In short, the sexes got started when one group of organisms (females) began to specialize in competing among themselves for resources, while another group (males) specialized in competing among themselves for access to these stockpiling organisms.[6]

This fusion of cells prior to division originally meant little more than an opportunity for small cells to proliferate—cells that otherwise would have been too cytoplasmically impoverished to do so. At this primeval stage, females can be seen as the richer, hardier life forms, males as con artists preying upon them for survival. But soon thereafter the path away from cloning—simple duplication of the large cells—led literally to a whole new way of life: a bizarre, exploitative, inefficient creativity that, once embarked upon, became very nearly inescapable. For along with fusion of two nonidentical cells came the potential for enormous diversity and adaptability. Sexual recombination provided a hedge against extinction in the see-saw world of environmental fluctuations and competition between species for limited resources. George Williams, a biologist with a gift for analogies, has compared the offspring of sexual unions to so many different lottery tickets, whereas cloned offspring are simply photocopies of the same ticket.

Clearly, the gambler with a number of different tickets is the one most likely to hold the winner. The jackpot in this instance is the opportunity to leave a greater number of descendants. In evolutionary terms, the only pay-off that really matters is better-than-average representation in the gene pool of succeeding generations.[7]

Natural selection is the process by which genes become disproportionately represented in the population. Individuals differ in their abilities to evade predators and keep healthy, to survive and reproduce. They may be in competition with one another for food and shelter. But often there is more to reproduction than simply surviving long enough to reach breeding age, for individuals of one sex may have to compete among themselves for access to the other sex. The unsuccessful competitor in such contests suffers not death but few or no offspring. When differential reproduction hinges on intrasexual competition—typically competition between males for access to females—Darwin suggested that the term "sexual selection" be used to distinguish this struggle to reproduce from the more general process of the struggle for survival. In a recent refinement of Darwin's hundred-year-old theory of sexual selection, Robert Trivers has pointed out that inevitably the sex contributing most to the production and rearing of offspring is the sex that gets competed for. The investing sex becomes the limiting resource.[8]

From the outset, the sex investing most in offspring tended to be female. Among mammals, only the most unusual circumstances permit the male's contribution to offspring to rival the sum of energetic expenditures that motherhood requires: ovulation, gestation, and lactation. Hence, male competes with male to tap the reproductive resources of females.

Anisogamy was the initial inequality in cell size, and the legacy of anisogamy was a system in which males tended to compete for females. Success favored males with the wanderlust needed to find females, the aggressiveness and weaponry needed to defend access to them from competitors, and a promiscuity which permitted little pause in the pilgrimage from mate to mate. This, then, is the classic story, though we will turn shortly to some of the primate variations on this theme.

In species where females travel in groups and where, as a result, one male may be able to monopolize many females, the effects of competition among males are most pronounced. In these polygynous species, the stakes in the competition are very high since unsuccessful males never have an opportunity to breed at all. Therefore, male–male competition is at its most intense.

With such an increase in the intensity of competition comes an increase in evolutionary pressures selecting for aggressiveness, strength, and size among males. Consider the patas, a long-legged savanna-dwelling monkey who provides a fine example of what sexual selection can do. The chestnut-colored patas male, with his drooping white mustaches, is nearly twice the size of the female. A typical patas troop in Uganda or Senegal contains only one fully adult male, who more or less successfully controls reproductive access to seven or so adult females by preventing roving males from entering the troop. His energies are largely devoted to this defense of his harem and he plays a relatively small role—compared to primates who have just one mate—in rearing infants.

Different patas males rotate through this position of harem holder but the lifetime reproductive success of each will not necessarily be equal. Certain biographies give a male the advantage in leaving his genetic record: he who remains in control of a large harem a bit longer than average; he who takes over first one troop and then another; he who happens to find a harem in an isolated corner of the patas' range and who is by chance never usurped at all. Through a combination of luck and favorable genetic qualifications such as strength and size, such males leave more descendants.

In general, the greater the inequality of reproductive success among males of a species, the greater is the difference in size between males and females, as the patas case illustrates. By contrast, males and females are most similar in monogamous species where a single male is paired with a single female, both of whom defend their territory and nurture the young. In this situation, competition among males for mates is much reduced and therefore aggressiveness, strength, and size are no more advantageous to males than they are to females. This association of monogamy with equality in size and of polygyny with great inequality in size was noted as early as the nineteenth century by Theodore Gill (who studied monogamous seals) and by Darwin himself, in *The Descent of Man and Selection in Relation to Sex.* These correlations have stood the test of statistical analysis by more modern researchers with more and more reliable data.[9]

Differential investment in offspring, then, often associated with polygyny, creates powerful biases favoring differences between males and females. These biases are further magnified by the capacity of individuals to *choose* which members of the opposite sex they will mate with. For the limiting (female) sex is rarely composed of passive individuals. As Darwin pointed out, competition among males is complicated by the predilections of females. Among primates in the wild, female preferences may be critical. Except for humans and an occasional ape, breeding among primates is initiated by the female. Presumably, if a trait such as large body size helps a male to compete successfully against his rivals, it would not be to the advantage of females to mate with small males, approximately half of whose offspring would be sons who would have to compete with other males. Over time, females that selected large males would have a breeding advantage over those that did not, because their large sons would have the edge over smaller males.

But the virtues of large size are not limitless. Even though most bulls don't live in china shops, there are, nevertheless, costs attached to that much bulk. Limitations to male size include availability of food and the restrictions of gravity. Orangutans are among the most arboreal of apes; yet, a fully grown male (weighing up to 165 pounds) may become so large that the forest canopy no longer supports his weight and he is forced to travel long distances by walking along the tangled, leech-infested forest floor. Larger than the female by 25 to 50 percent, the male orangutan is confined by his foraging needs to a nearly solitary existence. Slowly, persistently, endlessly, the shaggy, phlegmatic red titan consumes vast quantities of unripe fruits and mature leaves, the junk foods left by more discriminating females. The female orang, because she is smaller, can afford to be a picky eater, selecting the nutritious shoots of new leaves and the ripest fruits.[10] Partly as a consequence of dietary differences, the male travels farther than the female, utilizing a much larger larder. The two sexes travel separately through the jungle. Occasionally their paths intersect and they may rendezvous to breed; more often they simply pass by like two ships in the night. The common complaint of the primatologist intrepid enough to study orangutans is that in hours and hours of observation the adults almost never do anything, almost never meet anyone; rather, they munch endlessly and rest.

ANISOGAMY and sexual selection have had profound implications for the evolution of all mammals, including primates. Gross size differences between males and females among patas monkeys and orangutans are just two cases from a multitude of possible examples. But there have also been some rather extraordinary developments in the primate order which complicate the picture. Foremost among these is the trend among primates for greater investment in each infant. In line with the general course of mammalian evolution, much of the cost for these babies falls upon mothers—but not all of it. For male involvement with offspring has been greatly altered and elaborated in this order.

Among mammals generally, primate males stand out for the unusually well-developed role they play in rearing offspring, in protecting, carrying, even in a few cases feeding them. Male primates are not just fighting machines, built to conquer other males and reap their rewards in copulations. Were this the case, the primate female would be a different creature, vastly more submissive, less assertive, and less complex than the creature who will be described in Chapter 7. The trend toward costly infants has led to unusually complicated relations between males and females. I will postpone discussion of the special role of primate males until Chapter 5, and focus here only on the first of the two interconnected trends, the in-

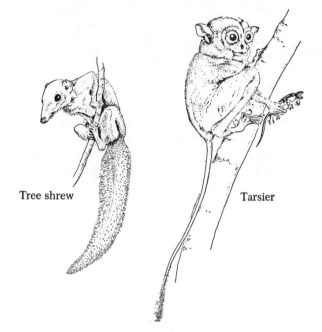

Tree shrew   Tarsier

creased investment in infants by primate mothers.

Reflect for a moment on the controversial tree shrew, the most primitive of living primates (if it is one, for experts disagree). This scurrying, squirrel-like animal provides the best extant model for a creature intermediate between the ancient insectivores and the earliest primates, which are represented today by such prosimians as the lemurs, lorises, and tarsiers. What distinguishes the tree shrew from all other primates is the remarkably tenuous thread between mother and offspring. Wild shrews live in holes in the ground or in the hollows of trees. The shy inhabitants of these hideaways may retreat to them singly or in pairs. The male may help prepare the nest, but once the female produces her litter—usually two or three babies—the male either leaves or is driven away, and it is not known what role, if any, he plays in protecting his family. The mother herself exhibits a rabbit-like nonchalance toward her progeny by absenting herself from her babies for several days at a stretch. Every forty-eight hours or so she returns just long enough to squirt into the waiting mouths a stream of milk uniquely adapted by remarkably high levels of fat and protein to compensate for maternal absenteeism. Characteristically, the milk of humans and other primates is very watery, low in protein, low in fat, adapted for more or less continuous suckling.[11]

The contrast between the tree shrew and the sprightly little tarsier—another prosimian with some remarkably simian characteristics—brings into focus the general trend of primate mothers toward greater and greater investment in offspring. The tarsier combines simian and prosimian traits together with other specializations uniquely its own. The rat-sized, rat-tailed tarsier has made a name for itself by possessing two long and spindly tarsal (ankle) bones capable of propelling its owner in long trajectories between trees. Tarsiers ogle their twilight world (dawn and dusk are their hours) through bulging eyes, and flick bat-like membranous ears, to locate the buzz or rustle of their next meal. But tarsiers are scarcely your run-of-the-mill prosimian. Instead of sniffing through the naked hairless nose (or rhinarium) of the other primitive forms, the tarsier breathes through nostrils covered by the hairy skin typical of monkeys. Below this furry nose, the upper lip is no longer tethered to the gum, so the mouth is freed for a rather advanced display of emotions. Most importantly, tarsier fetuses are nurtured within a remarkably monkey-like placenta. Whereas other prosimians, like the lemurs and lorises, retain a double lining that separates the maternal blood from that of her offspring, the tarsier, along with all other simians (the monkeys and apes), has an intimate placenta which is bathed directly by the mother's blood. Only a single membrane prevents the two bloodstreams from mingling. Unlike many of the lemurs and lorises, which have retained the shrew-like pattern of producing multiple young, the tarsier resembles the higher primates who (with the single exception of the marmosets) typically produce one infant at a time. Each infant is relatively large compared to the mother, but the reduction in litter size never-theless permits the mother to carry her single large baby clinging to the fur of her abdomen.

With the appearance of a creature like the tarsier, some an-cestral simian embarked on the distinctly primate style of re-production, which is characterized by great investment in each infant, a venerable mammalian tradition carried still fur-ther in this order. A portion of this investment is parceled out *in utero,* but it is in the investment after birth that primates really begin to diverge from other mammals. The evolution of the primate order exhibits a remarkably clear trend toward longer periods of infant and juvenile dependence, which among humans and apes may extend even into adolescence. The costs from cradle to grave of a $100,000 American baby[12] may be viewed as the somewhat absurdly bloated manifesta-tion of a reproductive strategy with roots in the Mesozoic era. How did this pattern of heavy investment get started?

For creatures living in a more or less stable environment, the world can be a very competitive place. Over time, most available niches will become filled by families successfully adapted to exploit them. Although on the surface conditions may seem predictable and benign, there is in fact consider-able competition for resources. An individual or family must be fine-tuned to the task of making a living in order to sur-vive. Hence, the quality of each offspring produced becomes more important than sheer quantities of babies made.

Imagine, as a contrast, a pair of creatures that have been swept away during a storm and find themselves rafted to an appropriate clime, uninhabited by any competing member of their species. The optimum evolutionary strategy for these newcomers will be to proliferate as rapidly as possible and fill all possible niches with their progeny before competitors ar-rive (in ecological jargon this strategy is termed "*r*-selected"; *r* stands for the intrinsic rate of increase in a population and is a measure of how prolific it can be). Much later in evolu-tionary time, the habitat becomes saturated with the descen-dants of various colonizing species in the various forms that they have evolved as part of exploiting the virgin territory. Now a shift in strategy is called for. The environment is so competitive that only full-size, healthy adults are able to survive on their own in many of its niches and breed successfully. The obvious strategy for prospective parents is to produce a few high-quality offspring and to protect them until they are self-sufficient (a "*K*-selected" strategy, in the jargon, where *K* symbolizes the saturated carrying capacity of the environment).[13] For primates, this last approach has become the norm. Because of the long mammalian history of differential investment in offspring, the burden of bigger, longer-lasting, superior babies falls in large measure upon the female, who must give birth to, suckle, and carry about this "improved" product.

After eight months of gestation the newborn chimpanzee is carried externally by its mother for an additional six months, and for the next six years the young chimp continues to travel and sleep each night with its mother in the nest she builds for herself. Until a male is ten years old, he does not leave his mother for periods longer than a few days. Full maturity is not achieved before the age of 15. A daughter matures sooner, at age 13, but remains with her mother even longer than a son does. Some mother–daughter ties persist into adulthood, and the daughter's offspring may receive assistance from the grandmother. It is from imitating their mothers that young chimpanzees of both sexes learn what routes to take through the forest, what plants to eat and when to eat them, how to build a nest, and how to make minimal tools such as the stripped twigs that chimpanzees use to fish for termites. Occasionally a mother may teach by active intervention; if an infant is seen nibbling on an unfamiliar, possibly poisonous, insect, the mother might knock the bug away with her hand.[14] Chimp fathers play a relatively small role in rearing young, though if a male happens to be in the vicinity he may contribute to the survival of his own offspring and that of his relations by threatening away predators or by driving away marauding males from some other chimp community. At maturity, a father may help his sons or nephews to become established in the local male hierarchy.

Even among primates such as baboons or people, where male investment in the survival of offspring is substantial, the task of nurturing falls more heavily on the mother. In the course of the most detailed long-term study of monkeys ever undertaken, Jeanne Altmann has tried to quantify what the costs of mothering actually are—an exceedingly difficult task. The subjects of Altmann's scrutiny were a single troop of savanna baboons living on the dusty plains of Amboseli, Kenya. Among her findings was the fact that females with dependent infants tend to suffer higher mortality rates than do females in other reproductive stages. Altmann hypothesizes—with the support of considerable data—that mothers suffer from the pressure to stay fed while encumbered by a clinging infant, and the pressure to provide nutrients—proteins, minerals, and fats—necessary to manufacture milk for this demanding hitchhiker.[15] The same point is brought home for the human case by a simple nutritional statistic: an average mother loses about eight pounds of fat during each period of lactation.[16]

ELABORATE theories, some of which I have touched on here, have been developed to explain why males in many species evolved to be so large and aggressive. I have suggested that some of these forces—but not all—may have been mitigated in the primate case by the greater involvement of primate males in rearing infants. But we still have only a partial answer to the original question of why males are able to dominate females. After all, competition is not (as is commonly believed) limited to males. Females strenuously compete with other females for the resources needed to support themselves and their progeny. Wouldn't larger size help females compete with one another, just as it apparently aids males? Furthermore, there are other advantages to being a big female. Big mothers may be better at producing, protecting, and nourishing their offspring. An old wives' tale that turns out to be true is that tall women have fewer complications at delivery.[17] A mother seal, to take another example, is better able to produce rich milk if she is very large. If a mother needs to defend her offspring from other members of her species, or from predators, great size and strength would again tilt the balance in her favor. Size is also an advantage in the face of certain environmental conditions, such as cold. Female langurs from populations living at high altitudes in the chilly Himalayas appear to be larger than female langurs on the plains, while males (who are typically larger than females) remain the same size at both locations. As a result, sexual dimorphism between male and female langurs is less pronounced at high altitudes. Similarly, the relaxation of ecological constraints such as food scarcity, which might once have selected against large size in females, may also lead, by a slightly different route, to the reduction of dimorphism as females are selected to attain the same body size as males.

There is possibly no better example of females with a pressing reason to outweigh and outrank males than the spotted hyena. These ungainly African carnivores, slouches by day, turn into ferociously efficient predators at night. According to Hans Kruuk, the Dutch ethologist who spent over three years on the plains of the Serengeti studying them, hyenas live in large communities of up to 80 animals. Within these clans, females unambiguously dominate males. Moreover, the female is larger than the male (on average, 56 kilos to his 49) and physically almost indistinguishable right down to her genitalia. In what appears to be the finishing touch to an elaborate mimicry, the female's elongated clitoris looks like a penis and is equipped with a sham scrotum—two sacs filled with fibrous tissue. So confusing is this resemblance between the sexes that from the time of Aristotle, hyenas have been rumored to be hermaphroditic.

Several explanations have been offered, but the most likely explanation for the female's male-like genitalia is that they are a side effect of the high levels of fetal androgens (or "masculizing" hormones) needed to stimulate the growth of females bigger than males.[18] Kruuk hypothesizes that the incentive for females to grow so large derives from the threat of cannibalism. Hyena adults sometimes attack and eat each other, and infants are in the greatest jeopardy since they are small and relatively defenseless. The reproductive success of a mother hyena depends on her ability to keep some other hyena—whatever its sex—from devouring her offspring. Her carnivorous lifestyle would also mean that large size would be an advantage rather than disadvantage in feeding, since she could hold her own as animals jostle one another and vie for position at kills.

As we shall see in Chapters 5 and 6, many primates face similar pressures to "design" big females. These pressures include competition for resources and territory, and most especially the need to protect vulnerable offspring from conspecifics, that is, members of their same species. Infanticide is widespread among primates, and the burden of defending infants falls upon the mother. Yet, despite such advantages to bigness, only in rare cases have female primates evolved to be as big as males. Only among the lemurs and in isolated instances among the higher primates are females dominant to males. Only under one particular type of breeding system, monogamy, do we routinely find anything approaching equality between the sexes in either size or rights of access to preferred resources. Why should this be so? What factors underlie these special cases?

This chapter has taken a circuitous route, starting with a primeval condition over a billion years ago, with the dawn of sexual reproduction. Organisms were composed of a single cell, and females tended to be larger than males. But what began as a different endowment in the metabolic resources of males and females developed, with time, into a complex reproductive system. There evolved males and females with quite different attributes. Females still contributed the lion's share to every baby produced, and sometimes mothers remained the larger sex. But elsewhere the original inequality in size became reversed. Countless generations after the initial cytoplasmic hijacking that launched inseminators (sperm) and resource-rich receptacles (ova) upon separate routes, sexual selection imposed powerful pressures which favored large males. Instead of bigger mothers (an efficient system from the vantage point of the species' energy economy), sexual selection turned the tables upside down. In order to compete with other males for the metabolic resources marshaled by females, males were selected for large body size, strength, and aggressiveness. Bigger males became virtually—but not totally—predestined. In some—but not all—cases, larger body size permitted males to dominate females. The time has come to consider some of these exceptions.

Consider the human species in its primitive simplicity . . . Though a man does not brood like a pigeon, and though he has no milk to suckle the young, and must in this respect be classed with the quadrupeds, his children are feeble and helpless for so long a time, that mother and children could ill dispense with the father's affection and the care which results from it. JEAN-JACQUES ROUSSEAU, 1762

3

Monogamous Primates: A Special Case

Nowhere among the social primates are females accorded more permanently privileged positions than among the monogamous species. There are some 37 of them, remarkably similar in many ways. Most of these creatures are rare and endangered, little known even to scientists. This is one reason for examining the monogamous primates in some detail. There is also a second, more personal reason for a closer look at animals that live in mated pairs. Sociobiological analysis invites a cynical view of living creatures. The best antidote I know of is the purely aesthetic device of treating each visit to the zoo as a museum tour, to view each animal as a creation and a demonstration of Nature's ingenuity. In this respect, the monogamous primates rank as masterpieces. Bear with the details then, for these creatures will not long be with us. However, as we travel through the forests of South America, Africa, and Madagascar and the archipelagoes off Malaysia where these animals are found, keep in mind the central question: Why should monogamy have such profound consequences for the status of females? Why has such a mating system evolved at all, or, considering the question from another point of view, why hasn't monogamy evolved more often? Consider first what monogamy means, and where we find it.

Two animals that breed and remain together to rear offspring are considered monogamous. The term is used here loosely and applies to a range of cases, including both lifelong intimates (such as gibbons) or breeding pairs that spend much time in tactile contact (such as titi monkeys), as well as more casual couples such as tree shrews who breed together and occupy and defend the same territory (what else they may do together, if anything, is not known). Monogamy is not necessarily synonomous with an exclusive pair bond, and either partner may occasionally solicit an outsider. But for both partners, opportunities to mate with outsiders will be limited, and both sexes are far less promiscuous than either solitary or polygynous species. Hence, male confidence in the paternity of his mate's offspring will be correspondingly greater, and male investment in offspring among monogamous species tends to be relatively high.

By such criteria, more than 90 percent of birds are known to be monogamous, while fewer than 4 percent of mammals are thought to be.[1] External eggs, of course, and the fact that mother birds are no better equipped than males to incubate them or to feed the emergent hatchlings, contribute to a situation in which male birds are selected to help rear young. By contrast, the more usual strategy of male mammals is to compete for the female, inseminate her, then seek another. The female is left to produce and suckle the young on her own. The primates, however, differ from the majority of mammals in this respect.

Female primates, like most mammals, invest more in each infant than males do. But there is a substantial tendency for primate males to remain in the vicinity of mothers and infants and to play some role in either protection or care of infants. In all but the most solitary members of the primate order (and these are few) males play a role in the general protection of group members—and that of course includes infants. Male primates are very tolerant toward infants born in their group and in many cases play a significant role in carrying or assisting older infants. But these tendencies—including both direct investment in infants (by carrying or feeding them) and indirect investment (deferring to mothers and infants over food) are more pronounced in monogamous species, where male investment begins to approach that of females.

Monogamous breeding systems are four times more common among primates than they are in mammals generally (on the basis of current information). Out of the 200-odd species of primates, as many as 37 (possibly more), roughly 18 percent, live in breeding pairs in which male investment in offspring is substantial and focused exclusively on the offspring of a single female. Additional species are suspected of living in such family groups, but this suspicion is based on census data rather than actual behavioral observations. Among species known to be monogamous, at least three (and probably more) are only monogamous under some conditions (Table 1). Tree shrews are a case in point. A Japanese research team, the Kawamichis, captured and marked 117 tree shrews (*Tupaia glis*) near the city of Singapore. They found two distinct patterns: whereas for most of the animals the home range of a single adult male and female overlapped almost completely, and hence they were essentially "living together," a small portion of the males who had the opportunity to do so (2 out of 17) lived polygynously, their ranges extending over the home ranges of several females. Unfortunately, tree shrews are problematic because we do not have good information on what role (if any) males play in rearing infants. Nevertheless, tree shrews provide an excellent example of what is meant by "facultative monogamy": a breeding system which may be adopted under some conditions but not others.

Humans, of course, provide another example of facultative monogamy. Approximately 20 percent of human societies are monogamous, 80 percent polgynous. Within polygynous societies, however, some males may have only one wife, others no wife at all, and this is usually a matter of economic circumstances. Monogamy in humans may be a matter of ecological necessity, or it may be imposed by legal sanctions. In any event, some of the general features of primate monogamy (like high male investment in infants) apply to the human case, but other, more specific features do not. (To take an obvious example, monogamous members of *Homo sapiens* are the only pairing primates that are not also arboreal or forest-dwelling. All the other terrestrial primates are polygynous.) For these reasons, I defer discussion of the complicated

Table 1. A taxonomy of monogamous primates.

SUBORDER PROSIMIANS			
Family Indriidae	Family Lemuridae	Family Tarsiidae	Family Tupaiidae
indri	mongoose lemur[b]	spectral tarsier	tree shrew[b]
(avahi)[a]	(red-bellied lemur)	Horsfield's tarsier	
(diademed sifaka)	(gentle lemur)		
	(ruffed lemur)		

SUBORDER ANTHROPOIDEA
   New World Monkeys

Family Callitrichidae	Family Cebidae
marmosets (3 to 8 species)[c]	aotus (night monkey)
tamarins (10 to 25 species)[c]	paleheaded saki
pygmy marmoset	titis (2 or 3 species)
callimico	

   Old World Monkeys

Subfamily Cercopithecinae	Subfamily Colobinae
de Brazza's monkey[b]	Mentawei Island langur
	simakobu[b]

   Apes and Humans
      Family Hylobatidae ("the lesser apes")
         gibbons (8 species)
         siamang
      Family Hominidae
         humans[b]

a. Species in parentheses are suspected of being monogamous; see note 2 in this chapter.

b. Monogamy may be facultative in these species; that is, it occurs under some conditions but not others. The species may be found in either monogamous or polygynous arrangements, depending on environmental factors.

c. Taxonomists disagree among themselves concerning how many species should be designated. Where there is substantial disagreement, as in the case of the Callitrichidae, I use conservative estimates in the text so as not to artificially inflate the proportion of species which are monogamous. This choice may turn out to be misleading if monogamous animals are especially prone to speciation, and this possibility should be kept in mind.

human case until later, where I deal specifically with human sexuality (Chapter 8).

Monogamy in the primate order is not a particularly exclusive club. Monogamous breeding systems can be found in each of the major primate groups and are by no means confined to the higher primates. Monogamy crops up among such "primitive" forms as tree shrews, indri, lemurs, and tarsiers,[2] and it appears again and again among the radiation of New World monkeys whose ancestors must have split off from the

Afro-Asian branch of the primates at some point prior to the Oligocene, more than 34 million years ago. The recurrence of monogamy among New World monkeys has led John Eisenberg and others to speculate that perhaps the common ancestor of New World monkeys and today's prosimians was also predisposed to a monogamous lifestyle. If so, this would suggest that a tendency for sustained relationships between a male and a particular female and a tendency for that male to involve himself in the rearing of offspring have a long heritage in the primate order, and that this tendency has "preadapted" females in different species to reschedule or reallocate investment in their offspring. Intensive care from two adults would permit twins to be born more frequently, or birth intervals to become shorter, since greater postnatal care from the father would allow a reduction in prenatal investment on the part of the mother. Helpless newborns would be assured of sustained nurture from both parents, a situation sometimes thought to characterize early humans.

The principal competing hypothesis is that monogamy has evolved independently in different primate species in response to particular environmental pressures. But whatever the causes of monogamy[3]—and we may never know them all—striking similarities exist among species that are monogamous. Often the two sexes are nearly indistinguishable in size and general appearance. If males are bigger, it is not by much, and in a few cases the female is larger, as is true for the common marmoset. Whereas pronounced size differences between males and females (sexual dimorphism) and distinctively large male canine teeth are characteristic of animals living in polygynous breeding systems, monomorphism (or same-size sexes) is typical of monogamous ones. In monogamous species, males defend the territory against male intruders, but females play an equivalent role in repulsing female intruders. Since males are not competing among themselves for harems, and since both sexes must compete for territories and resources, there is little selection pressure for a larger male sex endowed with uniquely male weaponry.

Dominance (defined in this instance as the ability of one animal to displace another from a resource both of them

want) is not, by and large, an issue between mates. Such interactions are rare. When they do occur, the female holds her
own against the male or actually dominates him. The high status of monogamous females is manifested in many other ways.
Males groom females more than females groom males (the re-
verse is true for most other primates). Females may lead the
family from one feeding location to another, and females have
priority at food sources. For the most part, aggression is
directed toward outsiders of the same sex; that is, females are
more aggressive toward other females, males toward males.

Lifestyles are remarkably similar among monogamous primates. Almost all are forest-dwelling and live in small ranges
or territories which both partners defend. Males invest in individual offspring either directly, by carrying or providing food
for them, or indirectly, by defending them or by yielding food
to the mother–offspring pair. In a few cases (such as wild titi
monkeys, captive tamarins, and gibbons), food is actually
shared among family members—a degree of "generosity"
which, apart from these monogamous primates, is usually
found only between mother and offspring, if at all.[4]

So MUCH for the bare bones of monogamous social
structures. To flesh out these sparse generalizations, let us
consider representative cases of monogamy from each of four
major evolutionary branches of the primate order: prosimians,
New World monkeys, Old World monkeys, and apes. On the
basis of sheer numbers (19 monogamous species), we might
begin with the New World monkeys, and in particular the
family Callitrichidae, which contains the marmosets, tamarins, and an odd-ball genus called *Callimico*, literally "beautiful little monkey," thought to be intermediate between the
Callitrichidae and the Cebidae, the other major group in the
New World branch of the primate family. (These little calli-
micos have only one baby at a time; among all other callitrichids, twinning is the rule.)

For many years, it was thought that the marmosets and
their kindred, fluffy little items with claws (they often look
more like Pekinese dogs than monkeys), must be primitive

forms retaining the ancient mammalian mode of producing young in litters. But recently Walter Leutenegger and other anatomists have questioned the view that callitrichid twinning is a primitive holdover. For one thing, the uterus of a marmoset has only a single chamber (technically it is "unicornuate"), a form suggesting that the immediate ancestors of marmosets produced one infant at a time, as other higher primates do. Twinning, then, is probably a secondary adaptation or specialization. But an adaptation for what? The answer is, apparently, fast reproduction within the limits of small body size.

Marmosets make their living by preying on insects or, when they are available, lizards, tree frogs, and small birds. They are also adroit consumers of fruits and sap. Trees in the territories of marmoset families are often leopard-spotted with holes bitten into the trunk by marmosets drilling for gum. Marmosets prosper along the edge of the forest and along streams, where insects are most abundant. All insect-eating primates are small, for they must survive on tiny packages of flying, scurrying protein. As ecological necessity selected for smaller and smaller females, mothers were confronted with the problem of delivering large and fast-maturing babies through proportionately small pelvises. Twinning evolved as a practical solution to this difficulty. Instead of bearing one enormous baby, mothers produced two or three infants that were merely large.[5]

The alternative—to produce fewer young—was not feasible within the framework of the marmoset's general ecological adaptation. Marmosets are primarily colonizing creatures. As soon as any disturbance or fluctuation in the environment opens an attractive niche, marmoset immigrants move quickly into the glade in the forest where the tree has fallen, into the patch that burned last year when lightning struck. Or else they proliferate along the forest edge where secondary growth is gradually yielding to jungle. Prosperity under such circumstances can be short-lived. To exploit environmental windfalls as they happen, callitrichids must be able to produce young rapidly. No other monkey is better at it: two at a time, as often as twice a year, while the sun shines.

But in hard times, as the forest closes in, the lizards and insects grow scarcer. Competition among marmosets for food becomes severe. Families retrench. Only the healthiest of a pair of twins survives. Birth spacing lengthens. Put another way, marmosets, like many small mammals, must be ecological double-agents, shuttling back and forth along a continuum ranging from *r*- to *K*-selected strategies—from reliance on proliferation to reliance on competition.[6]

The extraordinary reproductive capacity of the little marmoset is made possible by a breeding system that is as supportive to mothers as any in mammaldom. The system is maintained by the assertiveness of the breeding female, the deference shown to her by her mate and by nonbreeding members of the group, and most especially the behavior of the "father" (that is, the breeding male in the group who has a higher likelihood than most of being the progenitor). This male assumes primary responsibility for his offspring except when they are being suckled by the female.

This peculiar feature of marmosets was noted as early as the eighteenth century. Marmosets have scarcely been studied in their natural habitats in the forests of Central and South America, but they have been long and closely watched in captivity. Once a symbol of high fashion in France, they were worn as living ornaments that scurried in and out of billowing sleeves. The name *marmouset* derives from the Old French word for "grotesque little man." Across the channel in England marmosets were used as fashion accessories of another sort; rare creatures adorned the garden menageries of socialites. The midwife to the family of the Prince of Wales (later George IV) kept a particularly prosperous colony of marmosets, about whom the following observations were recorded in 1758:

> [The young] cling or stick very fast to the Breasts of their Dam; when they grow a little bigger, they hang to her Back or Shoulders who, when she is tired of them, will rub them off . . . [and] when she has quitted them, the Male immediately takes care of them, and suffers them to hang on his Back for a While to ease the Female.[7]

Small wonder that the female needs easing: infant marmosets and tamarins at birth weigh one-fifth—sometimes one-quar-

Golden lion tamarins

ter—as much as their mother! A female without a male to help would be sorely stressed; quite possibly she would fail to rear any young at all.

Immatures—who may be offspring of the breeding pair—continue to be tolerated in the territory; some remain even after they reach adulthood. One explanation may be that they too provide essential babysitting services. Infants may approach any older group member to beg for edible tidbits—a decapitated cricket or a lizard. Among golden lion tamarins—both the most glamorous and the most endangered of all the callitrichids—food sharing within the family apparently is adjusted to need. Adolescent tamarins give up food to younger group members at about the time the infants are being weaned, when the infants would be at maximum risk from malnutrition.[8]

Family life among tamarins and marmosets is remarkably serene for primates. The breeding female and her partner—who, although not dominant to his mate, is dominant to all other animals in the group—spend much of their time in proximity, huddling, grooming, or marking each other, and their territory, with scent. In many callitrichids, males feed their mates. Rarely is a partner out of sight or hearing. Typically, more than 90 percent of the female's copulations are with her mate. Fights between opposite-sex animals, particularly mates, are rarely observed. When fights do occur, it is usually in the context of a resident couple together chasing intruders out of their territory. In such cases, whichever partner is the same sex as the intruder invariably fights harder. The resident of the opposite sex may be more hospitable and may even attempt to mate with the outsider. This marmoset predilection for mating with outsiders is, of course, out of keeping with popular idealizations about "pair-bonding." (Among animals generally there is rarely a perfect conformity between individual behavior and the label that sums up the breeding system of that species; "monogamous" primates may seize opportunities to mate outside the family unit, just as "polygynous" primates sometimes form long-term consortships.)

Nevertheless, for some time keepers of callitrichids interested in maintaining high rates of reproduction have known that animals caged in pairs bred best. From the work of Gisela Epple, Rainier Lorenz, Harmut Rothe, and others it is known that the female—who is unabashedly dominant to the male—plays the major role in defending the integrity of the pair-bond by driving away strange females.[9] Almost all of the reports on the social behavior of marmosets in captivity stress the aggressiveness of these monkeys toward their rivals. Severe fighting between adults leading to injury and death has been reported for common marmosets, lion tamarins, black-faced tamarins, pinchés (a kind of tamarin), and even pygmy marmosets—the smallest and shyest of the New World monkeys, creatures no bigger than tennis balls. In all the marmosets studied, fights between females are both more frequent and fiercer than those between males. Female intolerance for other breeding females is a major deterrent to male philandering.

Recent studies of wild tamarins support the belief that a single breeding pair is the nucleus of the callitrichid family group, which may sometimes be surprisingly large. The largest reported group numbered 19 individuals; average size is closer to 7. These assemblages are apparently rather permeable, as subordinate animals wander in and out without much opposition.[10]

Why are transients tolerated? Possibly there is a mutual exchange of benefits going on: juveniles and other transients help to rear the offspring of territory holders and in return the transients share in a relatively safe feeding ground. An animal in a group is less vulnerable to predation than a solitary animal because a group maintains better reconnaissance and has a greater capacity to mob and successfully chase away a raptorial bird or other predator. From such relatively safe harbors, young tamarins and marmosets await the opening of a suitable territory where they themselves may pair and breed. In the meantime, they have also gained vital practice for their coming role as parents.

No matter how big the group, there is never more than a single breeding pair producing young in the case of callitrichids. These findings are in line with the discovery that among marmosets kept in captivity subordinate females do not ovulate. Apparently, it is the presence of the dominant female that somehow suppresses ovulation in these animals. If subordinate females are removed from the group and paired with adult males in cages of their own, or if the dominant female is removed, former subordinates begin to cycle and become pregnant almost immediately.[11] It is not known what would happen if a subordinate female did begin to breed. Would the dominant female drive her away? Or perhaps murder the subordinate's offspring, as wild dog and chimpanzee females are known to do?[12] In either of these events, it would behoove the subordinate to defer reproduction—which, after all, is a costly and risky enterprise—until she has a territory of her own. We will return in Chapter 6 to the question of why the presence of dominant females should be able to affect breeding in subordinates.

APART from the 15 or so callitrichids, some 22 other primate species breed in pairs. But monogamy has not arisen in the same way in all cases. At least 3 species—all Old World monkeys—appear to have adopted monogamous mating systems as a facultative response to predation by humans during historical times. In each case—the Mentawei Island langur (*Presbytis potenziani*), the rare simakobu monkey (*Nasalis concolor*), and the de Brazza's monkey (*Cercopithecus neglectus*)—the monogamous species is the only nonpolygynous member of its genus.

Both Mentawei langurs and simakobus are restricted to four islands off the western coast of Sumatra: Siberut, Sipora, North Pagai, and South Pagai. For the last several thousand years Mentawei Islanders have hunted them both for meat and for a tuft of hair to add to the lid of the hunter's quiver.[13] So rare is the simakobu that the only close-up photograph I have ever seen was one the zoologist Ron Tilson took of a wide-eyed but dead monkey staring up from a native stew; the pitiful creature had an uptilted nose resembling the juvenile form of its closest probable relative, the pinocchio-snouted proboscis monkeys of Borneo.

At the approach of hunters, the adult male Mentawei langur emits loud calls, bounding in wide circles, violently shaking branches in its path. The female and offspring cower motionless and silent in the treetops. Males among the West African de Brazza's monkeys may similarly play decoy—like the proverbial bird simulating a broken wing. If alerted in

Simakobu   De Brazza's monkey

time, however, both sexes hide. This is the simakobu strategy: to minimize movements and vocalizations and to freeze at danger. In all three cases monogamy is associated with "cryptic" behaviors, that is, hiding as a defense. Ron Tilson and Richard Tenaza, who first studied the Mentawei Island primates, and Annie Gautier-Hion, who first studied the de Brazza's monkey in Africa, have independently suggested that the shift to small family units among these beleaguered species may be a relatively recent adaptation to the pressures of being hunted.[14]

Many monogamous species have long-distance territorial calls of one sort or another. For that matter, so do all the langurs in the genus *Presbytis*. With one exception, only males give them. The exception is the single monogamous langur of the Mentawei Islands. Apparently, then, the Mentawei Island langurs have been monogamous long enough for a counterpart song to have evolved in females as well! There has also been enough time for sexual dimorphism to all but disappear. In contrast to most other langurs weighed, the Mentawei langur male (who averages 14.3 lbs.) is scarcely heavier than the female (average, 14.1 lbs.). Measuring from head to tail, the sexes are also about the same length.[15] In de Brazza's and simakobu monkeys, the other *nouveaux monogames*, however, males remain substantially larger than females. These are among the very few exceptions to the rule that among monogamous species, males and females are approximately the same size.

Before the report by Tilson and Tenaza (published in 1976), no one suspected that any Old World monkey—much less a langur—would be monogamous. Until then, known cases of monogamy in higher primates came from either New World monkeys or gibbons. My own first response was incredulity. It was only when Ron Tilson played me a recording of a male and a female Mentawei langur singing duets that I believed my ears. I had to. As has long been known for tropical birds, duetting is symptomatic of monogamy.

T ERRITORIAL calls range from combined howls
among the prosimian indri of Madagascar to the operatic complexity of songs by the gibbons and siamangs of Asia. In the
case of the gibbons, the female's contribution, aptly known as
her "great call," is the loudest and longest. The great call is repeated at regular intervals, either alternating with or against
a background of wailing male hoots. Similarly, the female's
acrobatics are typically more spectacular than the male's.

For these territorial displays to work, both sexes of gibbon
must validate their presence. A missing consort not only spoils
the duet but invites intrusion by another animal in search of a
mate: for territorial advertisements also serve as serenades.
Tenaza found that unmated male gibbons sang more than
mated ones, presumably to attract a partner.[16]

Every so often a brief paragraph in a scientific article will
hint at the allure of listening to wild gibbons. Joe and Elsie
Marshall, a couple stationed by SEATO in Malaysia, have combined an interest in opera with a love of nature and adventure
by taping gibbon songs throughout Southeast Asia. The Mar-
shalls recently wrote in *Science* that:

> Each pair of gibbons daily advertises its territory by loud singing
> accompanied by gymnastics—a show of force. The female's great
> call dominates the half-hour morning bout. It is a brilliant theme
> lasting 20 seconds or more, repeated every two to five minutes. It
> swells in volume after soft opening notes, achieves a climax in
> pitch, intensity or rapidity (at which time the gymnastics occur),
> then subsides. The male's short phrases, varying according to the
> species, either appear at appointed times during the great call,
> follow it as a coda, are interspersed between great calls (the fe-
> male's opening notes command his silence during her aria), or
> are broadcast from his sleeping tree during a predawn chorus.
> The male begins this chorus as a simple phrase after which he is
> silent for a quarter-minute while listening to his neighbors reply
> in kind. During the next 45 minutes or so until dawn, the male
> gradually adds to and embellishes the phrase to make it an elabo-
> rate, varying brief song. The female gibbon can utter the short
> calls of the mate; the male, however, never sings the great call.
> Subtle differences characterize individuals, and often the sub-
> adult joins the bout of great calls . . . The [female] Kloss' gib-
> bon's great call is probably the finest music uttered by a wild land
> mammal. Following the magnificent central trill is a slow, step-

wise descent in a low register . . . The fully elaborated predawn phrase of the male includes a trill . . . We heard these lovely sounds at 4 A.M. on a moonlit night from Tenaza's camp on South Pagai.[17]

These divas, the smallest of the apes—and, next to chimps, orangs, and gorillas, our closest primate relations—range throughout the islands and mainlands of Southeast Asia. The modern gibbon species are found swinging hands-over-head through the high canopy wherever deciduous monsoon and evergreen forests persist in a part of the world where tropical forest is being rapidly destroyed. They are the heirs of a large hylobatid ancestor with a cousinly resemblance to today's black siamang.

During the late Pleistocene period, the gibbon forebears swung their way across a land mass now called Sundaland, which extended from Southeast Asia to western Indonesia. Today, the area known as the Sunda Shelf is crisscrossed by shallow seaways that separate the Malaysian peninsula, the Malaysian archipelago, Borneo, Sumatra, Java, and the Mentawei Islands. In the Pleistocene, recurrent ice ages impounded the shallow seas, and for much of that period the Sunda Islands were connected to the mainland of Asia. Eight surviving species of gibbon and the siamang arose from the common Pleistocene ancestor. David Chivers, who has devoted the better part of ten years to the study and conservation of gibbons, estimated that as of 1977 there were around four million gibbons in the world. So rapidly is their universe being felled that the number will probably drop below 700,000 by 1982.[18]

The main division of the lesser apes is between the siamang and the gibbon. (The differences are profound enough that some taxonomists propose creating a separate genus, *Symphalangus*, just for the siamang.) Siamangs are bigger (nearly twice the size of most gibbons), uniquely endowed with an inflatable sac at the throat, and—by dint of this resonating instrument, which permits siamangs to bark, boom, and emit ear-shattering screams—noisier than any other ape. Siamangs live in smaller ranges than do gibbons but are casual about defending their borders. Siamangs are far less selective in their feeding patterns than are gibbons, consuming quantities of leaves (up to 50 percent of their diet) in addition to fruit, which is the staple diet of all gibbons. Finally, male siamangs are more directly attentive to infants than gibbon fathers are. From the end of the first year or so of the infant siamang's life, it is the male who carries it.

In contrast to siamangs, gibbons are selective feeders: 60 percent of their diet is fruit. Although much smaller than siamangs, gibbons need a territory nearly twice the size of that belonging to their leaf-eating black cousins to sustain their frugivorous habits. The effort that a siamang male spends carrying his offspring, a gibbon male diverts to defending his larder full of fruit trees. Gibbon males patrol their boundaries and, on rare occasions, physically assault trespassers, though usually defense is confined to more gentle pursuits such as singing.

Among the siamang and gibbons the two sexes are virtually indistinguishable in size and appearance, except for their genitals. Typically, both sexes are the same color (often black) with various pale circles and brow markings to enliven the face; some individuals may be either all buff or all black in color. But three species of gibbon—the pileated gibbon, the concolor gibbon, and the hoolock—are sexually dichromatic; that is, all males are colored one way and females quite another. The male is typically black, the female some paler shade, usually buff, light brown, or golden. If such a system does not seem odd, imagine a country in which all men are black with yellow hair, all women white with black hair.

Just why males and females in monogamous primates should look alike does not seem to me hard to understand. Freed to some extent from intrasexual competition, both sexes converge toward the optimum body type for their environment. It is a far more efficient system. Where extravagances occur, they are as likely to be useful in competition with other species as in competition with other members of the same sex; in such cases, both males and females should possess the trait. The mustache of the emperor tamarin is a case in point. Both male and female *Saquinus imperator* possess enormous drooping white mustaches—long enough when straightened to touch the shoulders! Why? In many parts of South America, several species of marmosets may frequent the same locale in overlapping ranges. Since their diets are similar, the neighboring species compete for food. Now although not particularly large, the emperor tamarins are extremely aggressive, and in zoos have to be kept apart from other species. In the wild, the emperors are just that: they dominate other marmosets and tamarins and chase them from fruiting trees and other transient resources. An obvious explanation for the extravagant mustaches, then, would be that they advertise before all other marmoset species the pugnacious character of their possessors, and accentuate his or her apparent size. Since male and female alike participate in defense of the larder, selection pressures to appear impressive would operate on both sexes.

Such arguments make it all the more surprising, then, that the two sexes would ever be different, particularly in monogamous species, and so it seems odder still that five of the seven sexually dichromatic primates are monogamous. Only one of several possible explanations for dichromatism strikes me as plausible, and it requires two assumptions: first, that at some point in the history of each dichromatic species, genes coding for a particular color pattern (say, buff) would become linked with superior female fitness (say, an inherited capacity to produce robust babies), and second, that males in these species had the ability to choose their mates. If so, a male who chose a buff-colored mate would leave more offspring than his non-discriminating neighbor, for in the monogamous situation a male's fitness is directly equivalent to that of his mate. Eventually, both males with a preference for buff females and buff females themselves would prevail in the population until only buff females were produced. This sort of sexual selection, involving genetic competition among females and male choice, would be the mirror image of the more common mammalian mode of male–male competition and choice by females. Because males and females in these species lead such similar lives, thrive at similar temperatures, eat similar foods, and have the same predators, only sexual selection seems to shed light on what would otherwise be a mystery.[19]

Whether or not gentlemen prefer blondes, males in some gibbon species appear to. But even though color is one of the more conspicuous characters animals may use in mate choice, there are probably other, equally important traits, particularly songs, which among all gibbons (except the sexually dichromatic hoolock gibbon) differ dramatically between the sexes, and differ to some extent from one individual to another. Only by becoming a gibbon, I expect, could we fully appreciate the sights, sounds, and smells that cause two wild and independent creatures to pair as they do. Suffice it to say there is more to the matter than just proximity. Choice is involved. And once a choice is made, the animals are apparently mated for life. A study of siamangs initiated by David Chivers has monitored the same pair for well over seven years. Although other monogamous primates have not been studied over a comparable time span, nothing has occurred yet to put the permanence of these unions in doubt. Nevertheless, it is true for gibbons, marmosets, and Mentawei langurs that partners who disappear are soon replaced.

Pairs who remain together may cooperate more efficiently in territorial defense and parenting. It is not known how much contact parents maintain with grown offspring, but certainly parental investment in some cases extends at least through first pairing. Tenaza has reported for Kloss's gibbon (yet another monogamous denizen of the Mentawei Islands, and the most musical of that songster genus *Hylobates*) that fathers assist sons in setting up territories adjacent to the parental home. The mother's contribution is a passive one: she keeps quiet, presumably to avoid discouraging an eligible female from a visit to the new territory.[20]

The form and timing of male investment varies from species to species, but in all cases the male's role in rearing offspring or in defending the territory needed to rear them is substantial. Because of such male investment, the question of paternity is of pressing importance to the male, who must devote himself exclusively to the rearing of particular infants. Were he to invest, mistakenly, in offspring sired by some other male, so far as posterity is concerned he would be squandering his life's work. Not surprisingly, males in monogamous species are exceedingly vigilant of breeding rights. The fine line between togetherness and surveillance, love and just plain jealousy, is epitomized by the endearing habits of titi monkeys (*Callicebus moloch*), New World monkeys who may sit side by side for hours upon a branch in some sylvan setting of Central or South America, a male and female with their long tails twined together.[21]

Barring adultery, both partners in a monogamous union invest heavily in the genetic fruits of their union. Reproductive success for the male is limited to the number of offspring his mate produces in her lifetime. She, in turn, is partially constrained by food resources available to her and the calories which she must divert to producing and suckling offspring. Under such circumstances, it is scarcely surprising that a male titi or marmoset might provision his offspring, or even his mate, or defer to them over food. By doing so, he contributes to the female's capacity to produce more offspring at shorter intervals. Indri are classically monogamous animals in this matter of male deference.

I NDRI is the largest living lemur. (A larger lemur, the size of a calf, became extinct in the seventeenth century.) Indri is the only lemur to lack a tail, and it certainly does not look much like a primate. Picture instead a burly black and white koala bear, somehow transported from its eucalyptus groves to the rainforests of Madagascar.

Picturing indri is one thing, seeing them another; they do not live in captivity. I perfected my own formula for finding indri several years ago on a visit to the east coast of Madagascar, the only place where indri exist: Enter a Malagasy rainforest and wander amid wet, mossy undergrowth and tree ferns, your eyes glued to the lichen-blotched canopy. Surprisingly, few birds will distract you on this island which, compared to the African mainland, is depauperate of birds. Orchids rather than birds draw your eyes from tree to tree. As you climb higher in the hilly terrain, you may enter thick forests of bamboo. At all altitudes, soft, black leeches will inch

Indri mother and infant

up your boots or brush onto your wrists. They are so tiny, with such a funny, inch-worm aspect that they scarcely seem serious until their rasp cuts through your skin. All the while, listen to the howls of indri which are here and there, like so many "damned, elusive Pimpernels," at the crest of a nearby hilltop, or just one valley over. As soon as a song sounds close, run as fast as possible along the nearest trail, and where the trail stops, throw yourself into the underbrush and thrash through it, being grateful that no poisonous snake ever managed to infiltrate the weird assemblage of creatures endemic to Madagascar. Deep in the thicket, above your head, you will see two adult indri, to all appearances identical. They may or may not have an offspring with them; if they do, only the mother will care for it. If they do not, you will be able to tell the female from the male only if you happen to see one urinate.

My own observations of indri were cursory. Only one person, Jonathan Pollock, has studied them in any depth.[22] Pollock put up with the leeches for a year, and by his own admission would like to return one day.

Indri live in a rainforest habitat so stable that resources are renewed only slowly. In contrast to the boom-or-bust world of the marmosets, these lemurs live at relatively constant low densities and reproduce slowly. Birth intervals are several years in duration. The population is presumably regulated by the availability of territories capable of supporting a pair of indri and their offspring from year to year. Fruit, but particularly leaf buds and new shoots—highly nutritious components of the indri diet—are in relatively short supply. Interestingly, Pollock found that, while foraging, indri females and infants were able to preempt the best feeding positions. Not only did males eat less on average, they ate fewer of the new shoots and other preferred food items. No other primate comes so close to fulfilling the Talmudic injunction, "Love your wife as you love yourself, and honor her more."

ALTHOUGH males subordinate themselves to females in the matter of food, it is the male indri who takes primary responsibility for troop defense by repelling invaders from the territory. In such encounters, female and offspring wait in some central portion of the range. Once again, then, monogamous primate males exhibit seemingly selfless sacrifice and a regard for female interests rarely found among polygynous species.

This brings us to the heart of the most fascinating puzzle about monogamy. What is in it for the male? He has the capacity to inseminate a dozen or more females; why should he focus on one to the exclusion of others?

It is possible, of course, that females mate selectively, accepting only monogamous males. But somehow this seems impractical. Fidelity would be too easy for a male to sham, only to abandon his mate once she was pregnant. It is more plausible to assume that the male himself is selected to stand by and assist the mother. If the survival rate of offspring fathered by helpful males were substantially higher than the survivorship of offspring whose fathers mated and left, males who stayed with their mates would on average have higher reproductive success.

The situation would be completely different, of course, if

the mother altered her own pattern of maternal investment. For example, if a marmoset mother slowed down her reproductive output, investing more in each infant as a chimpanzee mother does, the male would be freed for greater breeding opportunities elsewhere. Here, it seems to me, is the critical point: even at the risk of underestimating the complex coevolution between the sexes, it must be noted that by and large the reproductive strategy of the female is what determines how much the father must provide. Although many ecological pressures channel males and females among monogamous species toward greater interdependence and cooperation, we should not ignore those ways in which monogamy is imposed on males by females.

Female choice may be a factor in monogamy, but ultimately monogamy derives from constraints imposed by females at a more basic level. One way or another, females in these species are deploying themselves so that there is only one breeding female in each territory or group. Any prospect of polygyny would be precluded by fierce antagonism among females of breeding age. In most monogamous species, rival females are physically excluded from the territory by the aggressiveness of its mistress. In cases where the presence of other females is tolerated (as in wild tamarins), the integrity of the breeding unit is maintained by suppression of ovulation in subordinate females. Monogamy, then, is maintained by the way females deploy themselves both socially and geographically, and by female reproductive strategies which make male assistance (either in maintaining the territory or in actually rearing young) imperative. Neither sentiment nor sexuality enters the picture—until later.

Several scientists have asked themselves in recent years exactly what conditions predispose females to spread themselves singly through the environment, as opposed to living in groups, and in particular, what makes a breeding female in these species so intolerant of other breeding females? One possibility, for example, is that monogamy among primates is a by-product of life in stable habitats where animals are dispersed at low density throughout an area of continuously but slowly renewing resources. (Picture trees coming into fruit one after another in a tropical rainforest which has no seasons.) Jonathan Pollock described this situation for indri. Local fauna live close to the carrying capacity of the environment; life is competitive. To assure their own larder, each group must defend from its neighbors a specific territory full of the trees it feeds upon. The area that can be successfully defended against outsiders contains only enough food for two adults and their offspring. Asian gibbons, Malagasy indri, and the South American titi monkey all appear to fit this pattern.

Two adults can defend an area better than one, but why not two females? Especially if females are about the same size as males, and equally aggressive. But, that particular arrangement—two females, perhaps sisters, living together in a territory occasionally visited by a transient male whom they drive away once they are impregnated—has never been reported for primates. Yet it seems like a reasonable solution to the problem. The reason that no sororal system ever took off is probably that, quite simply, a father makes a better parent than an aunt does. A father shares half of his genes with his offspring and thus is more amenable to sacrifices on their behalf than is their aunt, who shares only one quarter of their genes. Whereas a male might be expected to defer to his mate over food and to make sacrifices in defense of mother and young, another breeding female would not.

An argument related to Pollock's, but one that does not require monogamous species to be territorial, is that particularly difficult physical conditions (such as food scarcity) select for female–female intolerance but nevertheless make it advantageous for a mother to have help from another adult in rearing her family.[23] It is extremely awkward, however, to specify what constitutes a "difficult" environment. By definition, any wild creature that sustains itself and breeds is living in a suitable environment. Primatologists can argue endlessly among themselves about whether certain monogamous primates qualify as occupants of a "poor primate's niche" or whether or not these creatures are making do with "suboptimal" resources. During the dry season, for example, mongoose lemurs literally eat like birds—hummingbirds. They subsist on the nectar of flowers.[24] Students of bats and birds, however, would tell us that nectar is a very rich source of food—scarcely a case of dietary slumming. But if nectar is such a choice food for primates, why do so few species specialize in eating it? And so the arguments would go, deep into the night. Callitrichids exhibit a similarly unconventional (for a primate) reliance on sap and in some cases nectar.

Most intriguingly, several species that have reverted to a nocturnal lifestyle after long stints as diurnal creatures are monogamous. They include *Aotus* (the only nocturnal anthropoid), *Avahi* (the only nocturnal Indriidae), and *Lemur mongoz* (the only nocturnal lemur). Not much would be known about the only nocturnal monkey *Aotus trivirgatus* in the Peruvian wilds where they live were it not for a young New Yorker named Pat Wright who fell in love with an *Aotus* named Flower that she found in a pet shop. Wright discovered that when she released Flower to wander near their summer home on Cape Cod, the saucer-eyed monkey always came back to her bed to roost. Wright felt that if she could just locate the monkey's sleeping lodge in the wild, she would be able to follow a group, in spite of the difficulties of working in the jungle at night. Based on Wright's recent fieldwork in Peru, we now recognize a number of similarities between *Aotus* and one of the partially nocturnal lemurs of Madagascar, *Lemur mongoz*.[25] Both the night monkey and the occasionally monogamous mongoose lemur[26] move in small, overlapping home ranges while foraging on a limited variety of food resources. In contrast to almost all other monogamous primates studied, neither species defends a territory, possibly because the resources they utilize are too scattered, limited, or unpredictable to efficiently defend. While not conclusive support for the hypothesis that monogamy is an adaptation to adversity, field studies of *Aotus* and mongoose lemurs suggest that these species are living in a larder of bread crumbs rather than slices, which two wives would not gracefully share.

GALLANT deference to females, exalted duets, the intimacies of tail twining, and supreme altruism in defense of one's mate—are these nothing more than strategies to cope with hard times? Is monogamy imposed on males by the needs and mutual intolerance of mothers coping with scant resources and big babies? These generalizations, which I regard as fairly accurate, do seem cynical. The stories of five primates that are *not* monogamous—polygynous species in which females nevertheless are unusually dominant—do little to romanticize the tale.

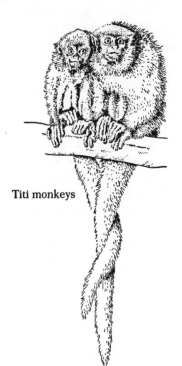

Titi monkeys

[*Odysseus as he sheds his beggar's disguise in the Great Hall and prepares to fight his rivals:*] *The hour has come to cook their lordships' mutton.*

HOMER, eighth century B.C. (translated by Robert Fitzgerald)

4

A Climate for Dominant Females

Chauvinists of both sexes have dipped into the primate literature to document their positions. Rarely are they disappointed. So diverse an order are the primates that it is a simple matter, by focusing on baboons, langurs, or orangutans, to "demonstrate" that male dominance is the natural state. By concentrating on prosimians, one can argue that female dominance is the primitive and basic condition, for among all the social lemurs ever studied, this is so.

Primates live in pairs, harems, unisex bands, multimale troops, as solitaries, as flexible communities that group and split, and as small subunits which attach to and disengage from very large associations. Females can be dominant, subordinate, equal, or not interested. Virtually every known social system, except polyandry (one female, several males) is represented. For just this reason, it is essential not to look at isolated cases but to examine the entirety of the evidence and to discern the patterns in it. Under what circumstances do males defer to females? The previous chapter dealt with monogamous primates in which females are accorded high status. This chapter focuses on the natural history of seven polygynous species in which males who are not confined to a single mate nevertheless defer to females during either part or all of the year. To a greater or lesser extent, females in these species take priority at feeding sites and control social access to other group members. An offending male who comes too close to a female or her infant is cuffed on the face or chased away, and in some cases males are relegated to the outskirts of the troop. These species stand in contrast to television narratives about "central male hierarchies" and "dominant male leaders."

A male in a polygynous species has the opportunity to inseminate many females, provided he is well-fed, healthy, competitive, and lucky. One would think that the male's own physical condition should be more important to him than that of any single female in his harem. In the majority of cases, this is true, and the male responds accordingly by behaving as a bully. Under what circumstances, then, is it adaptive for males in polygynous species to defer to females? How do social systems characterized by female dominance develop in a polygynous setting?

For two reasons, the natural starting point is with prosimians. These are the most primitive of living primates, and, more than any other, this suborder exhibits pronounced "matriarchal" tendencies. Female dominance typifies the group-living prosimians. In every social lemuroid that has been studied for one year or more—ring-tailed lemurs (*Lemur catta*), brown lemurs (*L. fulvus*), and white sifakas (*Propithecus verreauxi*)—females are known to dominate males with considerable frequency. Females also appear often to dominate males among the less well-studied ruffed lemurs (*L. variegatus*) and black ones (*L. macaco*) and may also do so among the beautiful but very rare diademed sifaka (*Propithecus diadema*). Indri, of course, provide a classic case of a species with high-status females, but their case (as with some of the other prosimians) is complicated by monogamy.

Several explanations have been offered for the high status of prosimian females. One is that female dominance is a phylogenetically ancient trait of the primate order, built into the history of the order, the "basic state," as some feminists might put it. This viewpoint is reminiscent of theories current in the nineteenth century concerning a "matriarchal" stage in human evolution, and it leads to the old conclusion that the basic primate condition has disappeared among all but a few primitive forms isolated on Madagascar. The alternative hypothesis, that the high status of females may be due to some feature of the lemurs' adaptation to life on this unusual island in the Indian Ocean, seems more promising.

The next step, then, is to consider the range of socioecological adaptations found not just among lemurs but among all known cases of polygynous primates whose females are unusually high ranking relative to males. These include the little talapoin monkey, also known as the "dwarf guenon" (*Miopithecus talapoin*) of Africa, and two species of squirrel monkeys (*Saimiri sciureus* and *S. oerstedii*), denizens of Central and South America which can be considered "Amazonian" in both senses of the word. The taxonomy of these species is outlined in Table 2.

In my opinion, the extraordinary parallels among the Old World talapoin, the New World squirrel monkey, and the array of social lemurs isolated on Madagascar make a fairly

Table 2. Polygynous primates in which females are thought to dominate males.

SUBORDER PROSIMIANS
Family Indriidae
sifakas (2 species)[a]
Family Lemuridae
ring-tailed lemur
black lemur
SUBORDER ANTHROPOIDEA
New World Monkeys
Family Cebidae
squirrel monkeys (2 species)[a]
Old World Monkeys
Subfamily Cercopithecinae
talapoin

a. In the case of both *Propithecus* (the sifakas) and *Saimiri* (the squirrel monkeys), one species of each genus (*Propithecus verreauxi* and *Saimiri sciureus*) is relatively well known, the other (*P. diadema* and *S. oerstedii*) virtually unknown. On the basis of superficial similarities between both members of the genus, I extrapolate to the genus as a whole. This is common practice in primatology, but readers should be aware that I am doing it.

strong (but not yet conclusive) case for attributing female dominance in these species to particular environmental pressures. For this reason, I will briefly summarize how the ring-tailed lemur, sifaka, squirrel monkey, and talapoin make their livings, and what these otherwise quite disparate creatures have in common.

IN 1962, at a tiny nature preserve called Berenty—a forested refuge for lemurs in the midst of a vast sea of sisal plantations in southern Madagascar—Alison Jolly completed the first detailed study of social behavior among prosimians. In particular, she observed sifaka and ring-tailed lemurs.

Lemur catta is probably the only primate anyone will ever mistake for a raccoon. The face of the ring-tailed lemur is white, with a bandit's mask of black about each eye; the tail is longer than the raccoon's and slightly less fluffy. While a troop of 20 or so lemurs walks along the ground, these black-and-white-striped tails are carried straight up in the air. When a typical band of, say, six adult males, nine females, and various immatures feeds in trees, the tails hang straight down. From a distance, the tails are so conspicuous that the first impression suggests a congress of floating fuzzy caterpillars. These tails have special significance for males, who use them in what Jolly terms "stink fights."

Male lemurs have two pairs of scent-producing organs, one in their armpits, the other on the inner surface of their forearms. Males prepare for stink fights by rubbing their tails over the glands of the forearm to anoint the tail with fatty secretions. Aggressive interactions between a male and another lemur usually begin with hard stares and proceed to tail waving, which scatters odor in the direction of the opponent. The typical outcome of such stink fights tells a tale of its own about how little the female lemur, as compared to the male, is impressed by such shenanigans. As Jolly describes it, "The male toward whom the tail waving is directed usually spats [gives a sharp squeak] and runs. A female often spats and cuffs the tail-waver."[1]

As Jolly interprets dominance relations among lemurs,

males are the posturers, more frequently engaged in threats and dominance interactions, but females are the seat of real power regarding resources:

> Dominant males swaggered, subordinates cringed . . . Females took little part in the threats . . . [Nevertheless] females were dominant over males, both in threats and in priority for food. Females at times bounced up to the dominant male and snatched a tamarind pod from his hand, cuffing him over the ear in the process. When females came into estrus, mating was not determined by the established threat hierarchy. Instead males fought around the females, slashing with their canines, and in three of four observed cases one subordinate won the battle. However, after mating he resumed his former low rank.[2]

Sifaka groups, which tend to be smaller than those of the ring-tails, often contain only three or four adult females, as many males, and various young. Sifaka resemble indri in physique, except that the former possess flowing white tails; these two relatively large, arboreal prosimians belong to the same family (Indriidae). Sifaka propel themselves vertically from tree to tree with powerful back legs. When forced to hop on the ground from one patch of trees to another, they do so awkwardly and with skittish vigilance. Like the ring-tail, the sifaka are sunbathers—or, according to local people, sun worshipers. Bare black skin on the chest and inner arms permits sifakas to absorb heat rapidly simply by opening their arms and gazing toward the heavens.

As with both indri and ring-tails, female sifakas (when they bother to interact with males) are dominant over males. Female sifakas routinely cuff males, usurp their feeding positions, and monopolize delicacies. But there is a twist to the dominance of females in this species: for a few weeks out of every year, in a flurry of breeding competition, male aggression dwarfs anything the females dish out, though almost all of this aggression is directed toward other males. During the breeding season the snowy fur of male sifakas is streaked with blood. Dominance positions occupied by males during the rest of the year can suddenly be reversed.[3] A subordinate male may become the dominant male, and vice versa, during the brief breeding season of a month or so.

Such extreme seasonality in breeding may be crucial for the evolution of the complex of traits that permits females in these polygynous primates to dominate males most of the time. All lemurs confine their breeding very strictly to a particular season and, interestingly, so do talapoin and squirrel monkeys. In this respect they differ from many other species of monkeys and apes (as well as humans), which, though they may have peaks in breeding, are nevertheless capable of conceiving throughout the year. In some cases a species (such as savanna baboons or Hanuman langurs) is a seasonal breeder in one location but not in another. In other cases, such as the rhesus macaque, the animals only breed—in fact may only be capable of breeding—at a certain period of the year; this is true for all rhesus that have been studied, whether they were in their natural habitat in India, in New Haven, or basking in the even Caribbean climate of Puerto Rico. Lemurs, saimiri, and talapoin resemble macaques in this respect, but males in these small-bodied species, with the costly metabolism that small size entails, have adopted quite an unusual strategy for dealing with the restriction on their opportunity to breed. Throughout most of the year they avoid the great expenditure of energy and the physical risks that maintaining high status costs a male. They opt out of the game, defer to females, and accept the status of "second-class citizens." When food is scarce, such a system contributes to the welfare of females and offspring, but it also permits males to save themselves (so to speak) and to play their cards all at once, when it counts most.

Squirrel monkeys provide the classic case of males who come into their own only briefly, during the eight weeks or so of the breeding season, when it most behooves them to do so. Among the smallest of the Cebidae (males and females average about 750 grams, under 2 pounds, only slightly larger than the biggest tamarin), these spooky-looking white-masked monkeys inhabit swampy to dry jungle throughout Central and South America. As with most New World primates, squirrel monkeys have scarcely been studied in their natural habitat. The best behavioral information for this species derives from the free-ranging denizens of Monkey Jungle, a large preserve just south of Miami. Frank DuMond has been observing the squirrel monkeys at Monkey Jungle since 1966. "From June to October," he writes, "the status of the adult male subgroup in the colony declines considerably and thus during the birth and infant-carrying time of year the colony becomes completely female dominated." Males are driven to the periphery of the group, presumably to reduce male interference in child rearing and foraging by females. Apart from the breeding season, males scarcely interact with female troop members. If males and females converge at the same time upon the same food source, males avoid females. If a strange male is experimentally released into the population at this time, it is the local females rather than males who drive him away. During the mating season, however, the squirrel monkey social order turns topsy-turvy. Come December, the Florida squirrel monkey's "pre-mating season," males undergo a remarkable change not only in behavior but, even more strikingly, in physique. DuMond calls this transformation the "fatted male phenomenon."[4]

Prior to the breeding season, squirrel monkey males build up subcutaneous fat deposits and are transformed from 715-gram weaklings into 937-gram bullies with the burly appearance of athletes in shoulder pads. Concurrently, the testicular regression characteristic of an off-season squirrel monkey is

Fatted male squirrel monkey Nonbreeding male

reversed, and males begin to produce active sperm. It is at this time of year that males compete with one another to establish a dominance hierarchy. Vying for status, males rush at and bump into one another, shriek, and engage in characteristic penile displays by which a dominant animal presses his opened thigh against the shoulder and head area of a crouching subordinate. The whole scene is reminiscent of carnival bumper cars operated by maniacal drivers assaulting one another with obscene gestures.

I T SEEMS likely that at least one species of Old World monkey, the talapoin of West Africa, has "sexual seasons" comparable to those of sifakas and squirrel monkeys.

Just as the squirrel monkey is the smallest of the Cebidae, so the talapoin is the tiniest of the Cercopithecinae, and in fact the smallest of all Old World monkeys. As in squirrel monkeys, males are slightly larger than females, but with considerable overlap between the two sexes so that it is difficult in the field to tell them apart. A female weighs just over one kilogram (2.5 pounds) and a male slightly more. The offspring, however, weigh a whopping 230 grams at birth—one-fifth of the mother's weight. Enormous babies also characterize squirrel monkeys and of course marmosets. These species differ from most other higher primates (but resemble prosimians) in the proportions of their "20-percent babies."

Talapoin monkeys have attracted considerable attention because in captivity (as noted by Thelma Rowell, A. F. Dixson and others) females routinely displace males. Unfortunately, very little is known about talapoins in their natural habitat. In the wild these small, olive-colored monkeys forage in mangrove swamps and thick forest for fruit and insects. As in squirrel monkeys, group size fluctuates, but large assemblages, up to 115 animals, have been counted in the wild by the French primatologist Annie Gautier-Hion. According to preliminary observations of talapoins in the wild by both Gautier-Hion and Rowell, a group might contain as many as 15 adult males, 27 females, and a number of immatures. But because these small and reclusive creatures occupy one of the hotter, more humid, and generally impenetrable parts of the world, the actual status of female talapoins relative to males is not yet well described. Annie Gautier-Hion has published data on a small sample of displacements involving food, positions, and access to receptive females. Since the data are lumped together for all types of displacements and for both the breeding and nonbreeding season, it is difficult to interpret them. Taking the field data at face value, one can conclude that although males are more likely to be involved in dominance interactions than females are, some adult females (it was not possible to identify individuals) are clearly displacing some adult males.[5] Such observations would be inconceivable among such species as langurs or hamadryas baboons. Even by the most conservative standards, one can say that talapoin are different from other polygynous monkeys, and there is no clear dominance of males over females.

Like squirrel monkeys and sifakas, talapoin monkeys are strictly seasonal breeders. Rowell reports that the prelude to this breeding season is a two- to three-week period of social hyperactivity among the males, who chase and threaten one another noisily. One interpretation of these observations would be that as among *Saimiri* and sifaka, males are fairly subdued for most of the year and concentrate their jostling for rank position into the breeding season. During and just prior to the breeding season, males become more aggressive toward any animal who challenges them, especially another male. For talapoin monkeys, this is still speculation, but if it is true that a male's rank is really important to him only during the breeding season, then a minor mystery about talapoin males would be cleared up: Although talapoin monkeys appear to differ from most other monkeys in the unusually high status of females, the pattern of play among juvenile talapoins is exactly the same as is found in other monkeys, with male juveniles being more aggressive.

It is not surprising that in the polygynous primates studied so far, behavioral differences between males and females, including dominance patterns, appear rather early. These are particularly manifested in play. Young males in species as diverse as rhesus macaques, patas monkeys, vervets, chimpanzees, orangutans, and humans are more aggressive, boisterous, and venturesome in their play than young females are. Females avoid much of the rough-and-tumble and keep their distance from the tag-playing males; they are less boisterous and exhibit a preference for affiliative behaviors. Because of this seeming "universal," Jaclyn Wolfheim decided to look at talapoins, for whom sex roles, so far as dominance was concerned, were reversed for captive adults. Yet, contrary to what might have been expected, Wolfheim found that sex differences in play behavior exactly paralleled those of other species where males remain more aggressive, and dominant, as adults. Wolfheim concluded that adult differences in behavior cannot always be accurately predicted from juvenile differences, a point that seems well demonstrated by her results.[6]

But could these results be cast in another light? There is fairly general agreement among animal behaviorists that play functions importantly in the acquisition and practice of skills that a young animal will need in adult life. Hence, males in many primates engage in mock fighting and chase games, females concentrate on what has been termed "play mothering"—carrying and holding infants born to adult females in the troop.[7] Since male talapoins as adults do not much engage in dominance and aggression, it was assumed that they had no need to practice such skills. But what about the breeding season? If, indeed, this is the brief moment of glory for aggressive males, talapoin males would profit in a reproductive sense every bit as much as baboons or macaques do from learning to display, bluff, and fight in the course of juvenile games.

Shared characteristics of talapoin and squirrel monkeys include small body size—which in turn correlates with high metabolic needs—relatively large babies, and seasonal breeding. Because these same features also characterize many prosimians, talapoins have long been considered as rather primitive, vestigial forms. The reader may remember that the same "accusation" was leveled at the marmosets because of their small size and habit of twinning. But as with marmosets, recent studies of the talapoin—particularly of its teeth—vindicate it as a thoroughly modern monkey.[8] Dentally, talapoins are similar to the Cercopithecinae, indicating that talapoins have evolved toward small size in the process of diverging from a common ancestor shared with the larger cercopithecines.

It seems likely that both the talapoin and members of the genus *Saimiri* are confined to small body sizes by recurrent seasonal scarcity of food and by the difficulties of making a living on fruits and insects in the face of competition with birds, bats, and other primates. Such small size and resulting high metabolism dictate an unusually tight energy budget for males. Given the small size, high metabolic needs, and seasonality of breeding, males would not only benefit from deferring all dominance activities to the one time of year when it matters, but they can, reproductively speaking, afford to. For the remainder of the year males can avoid risk and concentrate on staying fed—a period of competitive retrenchment comparable to Odysseus' brief stint disguised as a helpless beggar. When the moment arrives to regain Penelope, he sheds his camouflage and shows his muscle.

Several Malagasy lemurs resemble talapoin and saimiri in these respects. All are seasonal breeders. Hence, even in the case of lemurs, it is not clear whether the high status of females is due to phylogenetic inertia or to environmental factors such as those that molded talapoin and squirrel monkeys.

D ESPITE the problem of defining dominance, there is general agreement that in the majority of primates, males can usually displace females for access to commodities they both want. If pushed too far, females grimace and hold their ground. If a female, her infant, or even another relative is seriously threatened, she may retaliate by rushing at or chasing a male one-third again as big as she is. By and large, however, for primates such as baboons, gorillas, chimpanzees, and langurs, female deference to males is a way of life.

This is why the exceptions are so important. They demonstrate that primates are not totally locked into a pattern of male dominance. There are, as we have seen, two, perhaps three, important classes of exception. The first comprises monogamous primates living in family groups in which a male's reproductive success is more or less equivalent to that of his mate. The second includes seasonally breeding, small-bodied primates with high metabolic rates. Males of these species balance their metabolic budgets by strict conservation: through much of the year all dominance activities among males are suspended. Third, there are the prosimians; females in this suborder typically possess high status relative to males. In their case, however, we cannot be certain whether the ascendance of females is due to the evolutionary history of these most "primitive"of living primates, or to the same environmental pressures that weigh upon the other small, seasonal breeders.

Having raised the consciousness of primatologists, these exceptional species also proceed to raise again some sticky questions. Why are not dominant females more common among primates? Since females bear the greatest direct burden for the production of young, why don't females universally prevail over males at food sources? Why do females tolerate males at all? Like mythical Greek Amazons, and several real-life arboreal primates such as orangutans and tiny squirrel monkeys, a female could spend her life apart from males, permitting them access only once in a while, for breeding; she would eat better and possibly suffer less tribulation. Since there are typically fewer adult males than females, why don't females band together more often to assert their own interests? This question is particularly pressing in the face of a difficulty—a very big difficulty—female primates have with males: among a number of primates one of the most serious recurrent hazards that a mother has to face, comparable to predation and starvation, is the killing of her infant by males of her own species. Why, then, has she not evolved to sufficient size or dominance—as female hyenas have—to protect her own interests?

The first section of this book used a comparative approach to examine sexual equality and inequality in a variety of different species in the order primates. I now want to take particular problems in the lives of primates and examine how females with different evolutionary histories, existing under quite different environmental and social conditions, confront these challenges. There is of course no single composite of traits that constitute "the female primate." But there definitely are recurrent themes in the lives of these females—such as competition between females for resources, or female manipulation of males in order to protect offspring or elicit male investment. It is among these patterns that I believe we will find both answers to basic questions about the relations between males and females, and perhaps even answers to what are more important, and more elusive, questions about the nature (to the extent that it can ever be summarized at all) of female primates.

Ring-tailed lemurs

If you enlist all women in your cause and make them all abjure tyrannic Man, the obvious question then arises: "How-is this posterity to be provided?"   GILBERT AND SULLIVAN, 1884

5

The Pros and Cons of Males

In every well-studied species of group-living primates, males have been reported to offer some form of care or protection to infants. Whereas male participation in rearing offspring is commonplace among birds, it is fairly unusual for mammals generally—with a few striking exceptions such as African hunting dogs, where brothers help to feed pups born to females in their pack. The tendency of males to care for the young sets primates apart from the general mammalian herd.[1] Increasingly, anthropologists are beginning to focus on this peculiarity of our order to explain the emergence of *Homo sapiens*.[2] But along with clear-cut benefits of having males around as protectors, live-in babysitters, and even in a pinch as primary caretakers for infants old enough to survive without mother's milk, their presence also imposes a variety of costs.

Some are minor and easily discounted (for example, males take up space). Other costs are chronic and more serious: males usually compete with females and young for finite resources, and, because males are generally dominant, their privileged access to food could, in times of scarcity, erode the well-being of mothers and infants.[3]

It has been suggested that the cost of having males as competitors for food is offset by their assistance in defending the communal larder, guarding it from exploitation by neighboring groups. However, it should be noted that (with a few exceptions) females participate as much as males, or more, in defending feeding areas.[4] In fact, it could be argued that a troop needs males for territorial defense only because competing troops are also likely to have them. With males disbarred, troop defense would in all likelihood continue unabated—though perhaps more quietly—as an all-female enterprise.

Apart from the human case, there are very few examples of productive contributions by male primates to subsistence. Even where adult males hunt for game, as among chimpanzees, females and offspring receive little. Only among monogamous primates where males catch insects for their young (as do tamarins) is food sharing important. Furthermore, because troop members are often competing with one another for food, the presence of large males with large appetites is going to mean less for females and offspring. Compromise solutions have, of course, evolved: among orangutans and galagos, males travel apart from females; squirrel-monkey males forage on the outskirts of the troop; among indri, males defer to females over food. On the whole, however, females and offspring must simply grimace and bear the burden of having adult male mouths around, feeding.

Occasionally reports surface of a macaque or a langur troop traveling temporarily without an adult male, but such observations are very unusual. No one has ever reported a maleless baboon troop. It is doubtful whether such a troop could survive in the open country where baboons live, since males are needed to deter predators. This was brought home to me vividly one afternoon as I followed a troop of cynocephalus (or "dog-faced") baboons across the Amboseli plains of Kenya.[5]

With Mount Kilimanjaro in the background, baboons dotted the barren grassland, digging in the subsurface root system for nutritious corms. Animals thus intent on their work seem oblivious to danger, yet the likelihood of any predator approaching such a troop unaware is virtually zero. As the sun began to sink, the baboon troop moved toward a grove of fever trees where they would spend the night. Suddenly, sharp barks and convulsive "wahoo, wahoo" sounds from the male at the forefront alerted the entire group to the presence of a leopard skulking among fallen trees at the sleeping site. Every adult male in the troop save one (the alpha male, who remained as he was, escorting an estrous female) rushed forward and climbed into a dead tree overlooking the leopard's position. For nearly 40 minutes three of the males remained at their lookout, their brows puckered in intense concentration as they tracked the leopard's movements. Male baboons can and do sometimes lethally injure a leopard;[6] without the element of surprise, few leopards would take them on.

Most male primates are neither such stout defenders of the troop as baboons are, nor such absconded progenitors as tree shrews, galagos, and orangutans. Among macaque species (which, like baboons, tend to be found in large troops containing several adult males) the males safeguard the troop generally and also look out for particular infants born in the troop. Barbary macaque males (*Macaca sylvana*) assist selected infants through the harsh winters of the Atlas Mountains, carrying, cuddling, and warming them. During the birth season, adult male Japanese macaques move in from the periphery of the troop to carry about yearling infants whose mothers are otherwise occupied giving birth to that year's crop of new babies. In both baboons and macaques, male caretakers bias their attentions toward the offspring of particular females, often former consorts.[7]

Among those species where a single adult male does most of the breeding for as long as he is able to remain in the troop, there is typically little male-initiated contact between the male and infants. Males nonetheless do contribute to infant survival in manifold ways. Scattered through field reports for such species are accounts of daring and timely rescue: the patas male who pursued a jackal across the plains to retrieve from the predator an infant it had caught in its mouth; the langur male who shed aloofness long enough to tackle a raptor which had swooped down and nearly caught an infant in its talons; or, in one surprising example, a hamadryas baboon male who attended at the delivery of an infant. Perhaps no other single example illustrates so dramatically as this the existence of strong protective tendencies among male primates, even in a species notorious for the male's bossiness and possessive herding of females by biting them on the neck. A pregnant hamadryas baboon had gone into labor amid the steep Ethiopian cliffs near her sleeping site. As the baby's head emerged, the mother's rump protruded dangerously over a precipice. For a moment, the neonate dangled by the umbilical cord over the side of the cliff and would have fallen but for the male, who rushed forward, caught the falling infant, and handed it to its mother.[8]

Among gorillas, patas monkeys, and Hanuman langurs, males interact little with infants. Each species is notorious for paternal aloofness. Yet when the occasion arises, these same males play a crucial role in infant survival, even taking on the role of primary caretaker. Kelley and Alexander Harcourt, who studied wild gorillas among the Virunga volcanoes of Rwanda, have reported adoption of an orphan by the group's dominant or "silverback" male. In this case the apparent nonchalance of the gorilla male belies his critical role in emergencies: after its mother's death the infant sought succor not from other adult females but from the male who had previously ignored it. The infant traveled beside the male and even slept at night in the male's nest. My co-worker among langurs at Mount Abu, Jim Moore, observed a similar transformation from tolerant indifference to focused solicitude. In many thousands of hours of observations among langurs, this is the only such case either of us knows about; normally male solicitude for infants does not go beyond rescue from a passing predator or other transitory danger. In this instance, however, the male became personal guardian to an infant abandoned under duress by its mother. At the time of this writing—eight months later—the infant is still alive.

The langur male in question was probably a former troop leader who had been ousted by a new male (more about this process later). To avoid attacks on her infant by the usurping male, the mother left the troop and stayed near this male and other young males he was then traveling with. When the mother left, some weeks later, to rejoin her natal troop, her partially weaned infant son was left behind with the old male, who took over the main caretaking role, allowing the infant to huddle against him when it was cold. More than 90 percent of the infant's grooming interactions were with this male, who would carefully pick through the infant's fur.

The close relationship that developed between this langur male and the abandoned infant stands as a reminder to the situation-dependent character of relationships between males and infants. In retrospect, it seemed possible that this male, who had been in the mother's troop around the time the infant was conceived, was the father; certainly, he would be among the likely candidates. The theme of possible paternity will surface again and again in this and succeeding chapters, for the tendency among male primates to single out particular infants, watch over them, and even carry or otherwise care for them has a profound effect on relationships between males and females, and was a prominent feature on the social landscape in which female sexuality, as humans know it, first appeared.

On one hand, male primates direct special attention and care toward offspring likely to be their own; on the other hand, they sometimes exhibit brutal intolerance toward offspring sired by competitors. The propensity of males in a variety of primate species to kill offspring accompanying unfamiliar females is probably the most extreme cost females pay for having males around. An odd corollary of this is that in a number of species the most important contribution a male makes is protecting offspring born in his troop from attacks by other males. To tell this story, I will turn to the species I know best, *Presbytis entellus,* the Hanuman langur.

A TALE set in an unlikely scene: gray langur monkeys foraged peacefully along a shady lane leading to Abu's School for the Blind. The only sounds were the shuffling and crunching of monkeys eating dried plums, the cawing of Indian crows, and the tapping of white-tipped canes as two blind men clad in *dhotis* walked hand in hand up the tree-lined drive to the school.

In late July, at the monsoon's peak, the stately Bengal plums had provided all comers—birds, langurs, barefoot boys—with a sweet jamboree of burgundy-colored fruits. By September, only plums that had fallen to the ground remained. As usual, an aged female langur, whom I called Sol, fed on the periphery of the group, well apart from the other six adults, the five females, and the male that composed Hillside troop. Sol's black-gloved hands sifted nimbly through the leaf litter for remaindered jambol plums. When she found one, she turned her back on the others and privately gnawed at the husk with her worn teeth. Although she was the oldest female in the troop, Sol was subordinate to the others— nieces, daughters, and granddaughters—when it came to matters of food. If her feeding site caught the attention of others, some younger female would walk purposely toward Sol, mouth agape to expose pearly teeth—the universal vertebrate expression for "move away," which Sol never contested.

Only the infants had eyes for a world beyond the leaf litter covering the plums, an enthusiasm for the ebullient society of other little monkeys. The four infants included one black newborn, two cream yearlings, and Scratch, so named for the transitional state of his fur when I first saw him—black with scratches of white. Langurs at birth are covered with feathery black hair. The pale skin beneath is nearly transparent and the blood visible under the skin tints the furless face, ears, and bottoms of the infant's feet flamingo pink. By the third month the naked face has turned completely black, like an adult's; the black fur starts to turn white, beginning with the hair on top of the head and then somewhat later, with a little white goatee. This gradual transformation from dark to cream color continues until some six months after birth, when the infant has turned white as a polar bear. At adolescence, the coat changes once again to the regulation silver-gray of a Hanuman langur. Scratch had just turned cream.

Two young langurs on the ground in late afternoon constitute a quorum for sport; four virtually guarantee handstands, tail-swinging, and mad tag. But on this afternoon no playfellow strayed from its mother. Scratch strained for less staid company, but each time he tried to leave, his mother hauled him back by one rear leg. The absence of play was the only discernible flaw in the Eden of that late afternoon.

For months the infants had been in jeopardy: on numerous occasions in the last few weeks the male I knew as Mug, the alpha male of Hillside troop, had stalked infants in his recently acquired harem. Twice in this period Scratch had been wounded as Mug's sharp canines grazed his skin. This was the other side of the coin: the dark side of a primate male's ability to remember particular consorts and to single out for special attention infants likely to be (or in this case, unlikely to be) his own. Scratch had been conceived in August of the previous year when another male (Shifty Leftless) had just taken over Hillside troop.

In the previous assaults, Mug's attack on Scratch had been thwarted by old Sol and a second defender, the three-legged female Pawless. These females had intervened, throwing themselves at the male and wresting the infant from his grasp. But the odds lay in favor of Mug, who had the option to try, and try again, until he finally succeeded. So certain was Mug's eventual success that Scratch's mother had nearly ceased to resist—at least that is how I was tempted to interpret her lassitude.

Only days before, as the langurs fed in the swaying, flimsy branches of a jacaranda, his mother (I called her Itch) had allowed Scratch to tumble from the tree, 12 feet to the grass. The male had been the first to reach the fallen infant. Sol and Pawless arrived second, Itch last, risking nothing as she hung back from the desperate tussle that ensued. It was left to the older females, one of them a grandmother past the age of reproduction, to rescue the infant.

Itch herself was a young mother. Although Scratch was her second infant, she was still a few pounds short of her full adult body weight. With no conscious awareness of the fact— no more or less conscious, perhaps, than the self-absorption of a human teenager—she pursued selfish interests. She quested after status, the ripest fig, the most lucrative feeding position. A lifetime of breeding stretched before her if only she did not scotch the opportunity in a hopeless battle with some stronger animal, in this instance an 18-kilo langur male.

The assault beneath the jacaranda tree had been a near miss. With each failed attempt, Mug grew cagier. From the viewpoint of his victims, it made little difference whether this male's slyness was congenital or learned. The point was that his efforts were becoming more and more sophisticated. While looking everywhere but at the true focus of his attention, Mug would edge ever closer to his object, the infant Scratch. Every movement the male made, even innocent foraging, was a potential feint. The females were exhausted from having to stay ahead of him. The double duty of avoiding the male and still staying fed was taking its toll. To all appearances, Mug was master of this situation.

Mug, at about 12 years, was just entering his prime (an extremely lucky langur may live to be 30). A compact, muscular specimen, he was in most respects the model of a modern langur male: aloof, swaggering, memorably loud on a quotidian basis. Only the peafowl of the Indian hillsides give a call more haunting than the whoop-whaoop of a langur advertising his presence. Morning and evening, and occasionally in between, from the top of a tree, Mug would gaze skyward, fill his lungs with air and resonate. Males in surrounding troops whooped back. To underscore the point—"I'm here, hale and hearty, ready to fight you off!"—Mug would leap from branch to branch. Crashing of wood created special effects augmenting this spectacular audiovisual display. To all appearances, then, this fierce barrel-chested male was on top of his world. But in reality Mug was in a bit of a bind, the same bind that I believe has characterized langurs in this population for many generations.

This was Mount Abu, Rajasthan, in September of 1972.[9] Eight hundred years before, a twelfth-century sect of Jains built the Dilwara temple in these holiest of hills, geologically supposed to be the oldest in the world. Not far away, at Achalgarh, Shiva's toe is said to be enshrined. At the fountain of Gaumukh ("Cow's Mouth") below, the Rajput tribes claim to have originated. For centuries, pilgrims and honeymooners have paid homage at Abu's hilltop temples to Shiva, to Rama, and particularly to Lord Hanuman, monkey-servant to King Rama. En route they have fed, and still do feed, the local manifestations of Hanuman; for millennia, provisioning monkeys has been an alternate route to propitious reincarnation.

Each day during the tourist season, buses packed to bursting would wind their way from the parched plains, 4,000 feet up the forested mountainsides to the coolness of the hill station. Quantities of groundnuts and chickpeas were scattered among the langurs at the temples and bus stops by these would-be sowers of karmic harvest. What was the result? A burgeoning population of monkeys.

In the vicinity of the town and temples, where troops of langurs lived in fixed ranges inherited from mothers and grandmothers, there were 100 langurs per square mile, and the population was still growing.[10] Males of all ages who had been driven out of their natal troop by other males roamed the surrounding hillsides in nomadic all-male bands.

These fluctuating bands of males haunt the vicinity of troops containing both sexes and seek an opportunity to reenter. Competition between males for access to (or "possession" of) a harem is intense. Skirmishes between troop leaders and nomads are frequent and often result in injuries. Both-sex troops are under chronic siege, and the average male troop leader is able to hold on for only about two years. After an average of some 27 months, there comes a day when the bands of male intruders can no longer be beaten back, and the resident male is expelled. For a time a number of invaders may coexist in the troop, but when the dust settles, and hostilities cease, there is usually only a single male left in possession. The rest of the male band, together with the former leader and weaned male offspring, are banished from troop life until one of them can usurp control of the same or another troop. Unweaned offspring who remain with their mothers in the usurped troop may be killed by the new male.

Once installed, the usurper has little more than two years, give or take a few months, to guarantee that his genes will be represented in the next generation. If it were possible, the usurping male would be best served if the female's entire reproductive cycle were confined to his brief tenure. But the female cannot afford to cooperate with this compressed schedule of reproduction. To conceive, carry, give birth, and nurse the well-developed clinging infants that she must manufacture is a drain upon the mother. If her cycle were shorter, the cost might be so high that she would falter in the immediate and ever-present tasks of survival. So she spends six months and a half from conception to birth, another six or more months before weaning, and perhaps a few months beyond that before she conceives again.

The male can short-circuit the cycle, however. If he kills an infant sired by his predecessor, the mother almost immediately becomes sexually receptive again. Thus, in the act of removing the offspring of his genetic competitor, the new male also increases his own chance to procreate; the younger the baby, the more worthwhile (from the male's point of view) it is to kill it.

For Mug, time was running out; his eventual ouster was inevitable. If he was to beget an offspring with Scratch's mother, and have it survive, he would have to inseminate her soon. These were not Mug's thoughts, of course, but the conditions which had recurred in the lives of male langurs for generations. Males who responded to advantage enjoyed greater than average reproductive success.

At 6:30 P.M. on September 9, on the roof of the School for the Blind, Mug wrested Scratch from his mother and ran off with the infant in his mouth. Once again, the two older females, Sol and Pawless, charged the male to wrest the infant from him. Before they succeeded the infant was bitten in the

Hanuman langurs: Sol and Pawless charge Mug

skull and received a gash on his thigh and lower abdomen so deep that the intestines could be seen within. This is the only time in my career as a field primatologist that I have ever cried while making observations. Miraculously, Scratch survived, seemed to partially recover, and then disappeared.

MUG's behavior represents a straightforward solution to a major posterity problem. By eliminating a baby unlikely to be his own, the male reduces the number of offspring sired by his competitor; by inseminating the mother himself, he increases the ranks of his own progeny. But the male, seemingly in control, is in fact no less a prisoner in this system than females are. A male born or immigrating into an infanticidal population who failed to eliminate the offspring of his competitors would be at a reproductive disadvantage.[11] Both sexes, then, are trapped by the exigencies of natural selection, and have been since their origin.

If Mug's behavior seems unabashedly selfish, the mother's response is at first more puzzling. Why did Itch stand aside? Why did she not protect her infant? My only solution to the mystery of the "uninvolved" or "negligent" mother is hypothetical but, if correct, will lead directly back to the question raised earlier: Why do females fail to grow as large as males? What prevents females from standing up for themselves in the face of these warring polygynists? Why do females tolerate males at all?

To understand the problem we must consider barriers to cooperation among females which exist among primates generally. For that reason, before tackling the question of the role females as a sex play in their own subjugation, I need first to document that the enormous costs imposed on females by male reproductive strategies are in fact real, that they are not merely the product of heated sociobiological imaginations, and that such costs are not confined only to langurs; for there is by no means a consensus on this issue—even among primatologists.

The hypothesis that infanticide is a reproductive strategy favored by evolutionary forces and is adaptive for those individu-

als who succeed at it remains controversial. Powerful intellectual traditions in the social sciences hold that primate societies are integrated structures in which every animal cooper-
ates for the good of the group as a whole. Nowhere in this
scheme is there room for such unabashedly selfish behavior
as the killing of other males' offspring by a rival progenitor, to
promote his own reproductive success. From a group-oriented
perspective, so obviously detrimental a trait has to be mala-
daptive; infanticide must be regarded as a pathological aberration, brought about by environmental stress. Hence, various
authors have suggested that infanticidal langurs must have
been pushed beyond the limits of their "adaptive flexibility"
by crowding or human disturbance.[12]

Fueling controversy is the fact that in no species is infanticide a common, predictable, or easily observable event. Male
takeovers followed by attacks on infants, or missing infants,
have only been reported from 6 of the 13 locations where langurs have been studied.[13] In the roughly 20,000 hours that
langurs have now been studied, some 32 takeovers by invading males have been reported. In at least 20 of these cases,
takeovers were accompanied by the suspicious disappearance
of infants. Infanticide was assumed to have taken place on the
basis of circumstantial evidence: the timing of disappearance,
or repeated assaults on an infant by the new male in the day
or so before it disappeared. In only a handful of these cases
was the actual killing witnessed by a professional primatologist, although in areas where langurs live near human settle-
ments, eyewitness accounts by local people—sometimes reliable, sometimes not—have occasionally surfaced. The
following account from the Indian primatologist S. M. Mohnot, who has been studying desert-dwelling langurs near his
hometown of Jodhpur for over a decade, suggests why it is
that scientists—even when present—often miss witnessing
split-second attacks which leave infants wounded and missing
in their wake. The stunned Mohnot had scant warning:

> About 9:50 A.M. the male [who had recently taken control of the
> troop], with a sudden bound, was among the females. He
> grabbed the infant from the lap of [female Ti], clasped it in the
> right arm, held its left flank in his mouth and ran fast towards

the northern periphery of [the troop's home range]. The mother
. . . and two other females . . . rushed at the running male. The
mother twice blocked his passage but could not recapture the infant; the other two females also failed. All the while the infant
was screeching (cheen, cheen, cheen . . .). After running for
70–80 metres, the male stopped for a moment, took a quick bite
at the infant's left flank with his canines, dropped him on the
ground and sat near the bleeding infant. When the mother approached him, he barked loudly, tossed his head, bared his teeth
and stared at her . . . She was apparently frightened away . . . The
whole process, from the time the male snatched away the infant
and his subsequent dropping it, took 3 minutes.[14]

The dying infant was abandoned under a bush. Shortly
after, Mohnot witnessed the same male attack and kill two
other infants, in one case by slashing the infant's wind pipe
with his canines and in the other, biting is so deeply in the
thigh that the infant bled to death. But Mohnot's eyewitness
accounts are exceptional. More often, only a portion of the sequence of events surrounding infanticide will actually be
seen. Yukimaru Sugiyama, the Japanese scientist who first reported infanticide among langurs, was forced to reconstruct
events from partial evidence collected day by day. The following is taken from his field report of langurs living in South Indian teak forest near Dharwar:

On July 4, 1961, the invading male was seen attacking a
mother and infant; after repeated attempts, the male "at last
caught her from behind and bit the buttock [of the infant
which] stuck out from between her thighs." It was not until
the following day, however, that Sugiyama was able to get
close enough to the mother to examine the wounds, and not
until the second day following the attack that the infant disappeared. The body—probably carried off by scavenging birds—
was never found.[15]

A clear suspect with an obvious motive; a missing corpse;
and no eyewitnesses. Among humans, this is the stuff great
murder mysteries are made of. Among langurologists, it has
led to heated academic dispute. Hence, if I am going to examine the question of why female langurs do not better defend
their interests against infanticidal males, why they don't come
together in a united front to forestall infanticide, it is incum-

bent upon me to first demonstrate that there is indeed a need for females to do so!

Toward this end, I need to review the evidence: recurrent patterns of behavior among langurs at different sites, as well as among monkeys and apes in a wide range of species. Again and again, circumstantial evidence and eyewitness accounts converge and point to the invading male, the male entering the breeding system from outside it. Some 46 langur infants have disappeared just after a new male entered the troop. Hundreds of unsuccessful assaults on infants, including cases where the infant was wounded but did not die, have been seen when new males usurped troops. In thousands of hours of observation, normally tolerant male langurs have never been seen to attack an infant under any other circumstance. Confronted with a newcomer, females are at once hostile to the new male and intensely protective of infants. By contrast, females are nonchalant in the presence of an established troop leader, allowing their infants to play about him, to use his back as a trampoline, and to swing on his tail. At worst, a troop leader may grimace at a particularly obstreperous infant or threaten an infant interfering with him if he happens to be copulating. Most importantly, males aim damaging attacks at infants which are almost certainly not their own—that is, they attack only offspring accompanied by unfamiliar females. If infanticidal males are simply monkey psychopaths, with no method to their madness, one would not expect such fine distinctions to be made.

Compared with *Presbytis entellus,* other members of the colobine subfamily have scarcely been studied. Nevertheless, it is increasingly apparent that infanticide also occurs in other members of this widespread group of African and Asian leaf-eating monkeys. (Colobines are known as leaf eaters because they have the capacity to digest mature leaves. Many colobines prefer fruit, seeds, flowers, and even insects when available and are far more omnivorous than the term leaf eater suggests.) The first hint was a report from Rasanayagam Rudran, a young Sri Lankan naturalist who had been commuting between his home in Kandy, Sri Lanka, to nearby forests that abound with a cousin of the Hanuman langur, *Presbytis senex,* the "purple-faced leaf eater." Over a two-year period Rudran reported five takeovers among troops that he was intermittently visiting. After such takeovers, infants were badly mauled or missing altogether.[16] A few years later, in Malaysia, primatologists reported a similar pattern of male takeovers and missing infants among the silvered leaf monkeys (*Presbytis cristata*) of Kuala Selangor. In her most recent report on these monkeys, Kathy Wolf described the killing of an infant just after a rank reversal, when one male precipitously rose in rank above an injured rival:

> Bozo [the new alpha male] calmly approached the adult female and took the infant from her. It looked just like a normal infant transfer until the mother and Max [former leader] attacked Bozo. He turned and dropped the infant and when the mother retrieved it I could see that Bozo had slashed open the abdomen and [the intestines] were hanging out.[17]

The concurrence of male takeovers and either attacks on infants or their disappearance among such geographically distant colobines in India, Sri Lanka, Malaysia, and possibly also among the black-and-white colobus monkeys of Africa,[18] suggests that propensities for infanticidal behavior date far back in time, some millions of years when the split between the various species in the colobine subfamily took place. It seems possible that some constellation of factors leading to short male tenure among leaf-eating monkeys may make infanticide a particularly advantageous strategy for males in this subfamily.

Infanticide is not limited just to this leaf-eating branch of the Old World monkeys. Infant killing is now known to occur among the great apes, New World monkeys, and even cercopithecines—the other and very successful group that, together with colobines, makes up the Old World monkeys. A few years ago, I doubt that any primatologist would have taken seriously the possibility that infanticide is a serious threat to infant survival among cercopithecines. Now the work of Thomas Struhsaker and his coworkers in the Kibale forest of Uganda, and of Curt Busse and W. J. Hamilton in Botswana, forces us to reconsider.

From 1970 until the present, with little respite—even during the troubled period of Idi Amin's dictatorship and the war that ousted him—a team of primatologists under the direction of Thomas Struhsaker monitored eight primate species living in the remarkably rich rainforest of western Uganda, near the Mountains of the Moon. Although primate populations here are virtually undisturbed, and never provisioned, the population density of colobus monkeys living here is twice that of their colobine cousins the langurs living in provisioned areas such as Mount Abu. The total biomass of primates at Struhsaker's site was 2,200 kilos per 100 hectares! Among the main goals of the study was to learn about the life ways of the little-known "redtail" monkeys (*Cercopithecus ascanius*), comical creatures with long copper-colored tails, fly-away sideburns, and a heart-shaped smudge of white, smack in the middle of the nose. Struhsaker chose two study areas within the Kibale forest for his redtail study. Shortly after he started work at the second site, one of the redtail groups was taken over by an outside male that Struhsaker dubbed New Male. In his technical account of ensuing events, Struhsaker wrote:

> An infant was born . . . on the night or early morning of [December] 21st–22nd. No aggression was shown by New Male toward this new infant until 1640 hours of the 22nd when he suddenly attacked and killed this infant. Initially I saw a ball of screaming monkeys tumbling down a liana thicket to the ground. New Male then ran off along the ground with the newborn infant in his mouth while being chased by an adult female and several others. The adult female gave shrill shrieks and persisted in the chase longer than the others. The chase terminated within a few seconds and New Male climbed into a tree and began feeding on the dead infant. A monkey of adult female appearance sat 10 meters away chirping toward him. New Male chewed off and discarded pieces of the hind skull and then ate out the brains.[19]

This was the second of two killings by the incoming male. Both victims were new infants, presumably unrelated to the usurper. What impressed Struhsaker was not just murder and cannibalism among supposedly nonaggressive forest monkeys but the extraordinary coincidence that would permit him—in spite of the small number of hours these monkeys have been observed—to be present at the precise moment the killing was committed. (Imagine a police officer hoping to witness a murder in Detroit by singling out 70 individuals and watching them for a year. However high the actual murder rate, his chances would be slim.)

As Struhsaker put it: "The fact that male replacement followed by infanticide has been reported so rarely and from so few species is, I believe, more a statement of ignorance than fact. It seems plausible that with more longterm studies of Old World monkeys having one-male groups, we will find this phenomenon to be the rule rather than the exception." In 1977, the year these remarks were published, scientists at a French field station in the Ivory Coast, West Africa, witnessed infanticide by an incoming troop leader among another little known cercopithecine, *Cercopithecus cambelli lowei.*[20] The following year, Tom Butynski, working at the same study site as Struhsaker, reported infanticide in another scarcely studied denizen of the Kibale forest, *Cercopithecus mitis,* the beautiful "blue monkeys" whose elegant leaping forms—although hardly household visions today—were immortalized centuries ago on the frescoed Minoan walls at Thera, near Crete.

This new awareness that infanticide is a recurring problem among primates has led scientists to reinterpret well-known primate behaviors in a new light. Savanna baboons, for example, will frequently pick up and carry young infants when engaged in confrontations with other adult males, particularly incoming males of high rank. The old explanation was that a resident male who took up an infant was using it as a shield or an "agonistic buffer" to prevent the strange male from attacking him. But an accidental discovery among chacma baboons near the Okavango swamp in Botswana caused two scientists, Curt Busse and W. J. Hamilton, to consider a new interpretation. Busse and Hamilton were temporarily immobilizing baboons in order to obtain blood and other samples. To their chagrin, they learned that if a mother is drugged and unable to protect her infant, immigrant males may take advantage of this to kill her infant. Busse and Hamilton now hypothesize that males who are long-term residents, and former consorts of the mother, carry infants not to gain an advantage in the encounter (the old agonistic buffering hypothesis) but in order to protect from infanticide a possible relation.[21]

The history of our knowledge about primate infanticide is in many ways a parable for the biases and fallibility that plague observational sciences: we discount the unimaginable and fail to see what we do not expect. At precisely the same time that accumulating evidence from primate field studies was forcing scientists to reconsider the myth of primate gentleness, the notion of a behavioral continuum between humans and other primates became increasingly controversial. In the early days of ethology—what might be termed the Lorenzian era—humans were thought to be uniquely murderous; other animals—monkeys and apes among them—were supposed to be group-oriented species, civic-minded by comparison. Then, new findings generated powerful incentives to cast the primate evidence in a different light. Equally strong visceral barriers, however, prevented us from believing that such seemingly peaceable creatures (and such near relatives) could be so brutal. I believe that one of the important factors determining whether infanticide was recorded or missed may have simply been expectation.

The career of Rasanayagam Rudran, the young Singhalese primatologist who first reported infanticide among the purple-faced langurs, underscores this point. Having been starkly confronted by infant killing among purple-faced leaf monkeys in Sri Lanka years earlier, he was alert to the possibility of infanticide whenever infants disappeared at the time of a troop takeover among blue monkeys in the Kibale forest. Rudran was the first to suspect infanticide among these monkeys—though it was Butynski rather than Rudran who finally confirmed these suspicions. By the time Rudran went to South America to study howler monkeys (*Alouatta seniculus*) in Venezuela, he was on guard whenever a new male entered a breeding system from outside it. Back in the 1950s native observers had told earlier researchers that howler males sometimes killed infants, just as about the same time local Hindus were telling langurologists about infanticide. These reports were dismissed as folklore. Though the odds were against him, Rudran arranged his schedule to spend nearly all his time near a recently usurped troop in order to document infanticide in howler monkeys. Owing to this hunch and his special preparations, Rudran became the first to witness and film what otherwise would have been unrecorded killings. Rudran's report of infanticide among howler monkeys has now been confirmed several times by the zoologist Ranka Sekulic and others.[22] Furthermore, examination of old records for howler monkeys, Nilgiri langurs (*Presbytis johnii*), and chimpanzees suggests that infanticide may have been going on undetected at several sites.

The best documented of these cases is that of the Gombe Stream Reserve. David Bygott was following a band of five male chimpanzees when they encountered a strange female from another community. In hundreds of hours of observation Bygott had never seen her before. This female and her infant were immediately and intensely attacked by the males. For a few moments the screaming mass of chimps disappeared from Bygott's view. When he relocated them, the strange female was gone. One of the males held a struggling infant. "Its nose was bleeding as though from a blow and [the male], holding the infant's legs, intermittently beat its head against a branch. After three minutes, he began to eat the flesh from the thighs of the infant which stopped struggling and calling."[23]

As usual, the first response was to discount this gruesome incident as an aberration. But subsequent observations of adult males murdering and cannibalizing the offspring of alien groups by Japanese researchers at the Budongo Forest in Uganda and in the Mahale mountains of Tanzania, as well as reports by additional members of Jane Goodall's team at Gombe Stream Reserve, now lead to the conclusion that the behavior is widespread. In a recent summary of such reports of infant killing and cannibalism among chimpanzees, Jane Goodall describes four additional cases of infanticide by males and three cases involving adult females (discussed in Chapter 6) which have occurred at Gombe since Bygott's initial observations in 1971. In each case involving males, the victim belonged to a strange female and hence was unlikely to be a descendant of any of the murderers. Driven to find out whether infanticide was a recent trend or something that had been going on unnoticed all along, Goodall carefully examined her records for the past thirteen years: she came to suspect that the recent cases at Gombe were not the first.[24]

Such findings lead to the unsettling realization that the gentle souls we claim as our near relations are an extraordinarily murderous lot. Nowhere has this realization been harder to accept than in the case of the great apes, those ungainly, expressive anthropoids whose kisses, handshakes, and other antics most of us have warmed to at one time or another on the pages of *National Geographic* or on our television screens. Yet outside of the colobines, apes have the highest incidence of murder ever recorded for higher primates. Apparently most of these killings are infanticides, a particularly striking observation when the relatively small number of births among apes is taken into account. In some 11,000 hours of observation of fewer than 300 gorillas atop the cloud-enshrouded peaks of the Virunga volcanoes in Rwanda, Dian Fossey has witnessed three cases of adult males committing infanticide and she has inferred the occurrence of three others. However, infants are not the only victims of these seemingly affable giants: the crania of two adult males that Fossey examined had gorilla canines embedded in the bone. It is estimated by her that one-quarter of all gorilla mortality is due to injuries inflicted by other gorillas.[25] The annual reported murder rate of New York City—20.5 per 100,000—seems almost civilized by comparison. Under the circumstances it seems somewhat ironic that the old definition of "murder," which originally meant "to kill inhumanly," has been rather widely discarded and replaced by a definition that applies only to *Homo sapiens*, "the unlawful killing of another human being."

IN NO species is infanticide a common event. But what I hope to have demonstrated here is that for many primates it is, and has been throughout a substantial portion of their evolutionary history, a recurrent hazard. As with predation, infanticide does not occur often, but when it does take place there are drastic consequences for the fitness of the animals involved.

We may look at infanticide as one maneuver in a contest. To understand how the game works, though, we must bear in mind just who the players are. In one sense, of course, it is a competition between two individual males to see which of them can leave the most progeny. But looked at another way, it is a contest between all males and all females of the species. Any mother of an infant has nothing to gain from losing it, regardless of who the father is. Indeed, the stakes in this struggle between male and female are so high that it is hard to understand why large body size, big canines—a whole battery of traits to help defend themselves against males—have not evolved among females as well as their male opponents. Each primate infant is not only very vulnerable but extraordinarily costly to the female who produced it. If ever there were a biological incentive for females to be as big or as strong and aggressive as males, surely infanticide would be one.

A female strong enough to exclude males would have little use for them. She could at once obviate competition for food and forestall infanticide. A truce once a year or so would be sufficient to ensure breeding. Force aside, other avenues of defense are, in principle, also open to females. Since infanticide depends for its evolutionary feasibility on the timely insemination of the mother by the usurper who has killed her infant, females could simply fail to ovulate in the presence of an infanticidal male. To each creature, however weak, the option is there to vote, so to speak, with her genes.

Yet females—among langurs, for example—do not sexually boycott males with infanticidal tendencies. They do not selflessly ally themselves to fend off the male (a fully determined combination of two or several 11-kilo langur females would outweigh a single 18-kilo male). They do not grow as large as males. These facts all suggest that counterselection is at work. For the game of reproduction includes yet another opposition: female against female. Just as individual males compete with each other to leave the maximum number of progeny, so individual females are, willy-nilly, pitted against each other by the genetic rules. And any female who grew bigger, and therefore gained an advantage against the threat of a male's attack, would lose ground in the race with her sisters. Resources that otherwise would be available for reproduction would have to be diverted into attaining and maintaining a greater bulk. In particular, she would be spending food energy to make a bigger body for herself instead of more babies for posterity.

In any given generation, the balance of body size against reproductive potential is a delicate one, and the evolutionary problem is complicated by the fact that average body size increases very slowly in a population—ounce by ounce over many generations. Early in this process, a female who is only slightly larger than her sisters would experience none of the advantages of greater size, since she would still be pitted against considerably larger males. Droughts and food shortages, meanwhile, are recurrent hazards of the langur's environment; under these circumstances, the smaller the female, the better adapted she is at converting the available resources into offspring. The advantage to self-defense of growing bigger would be nonexistent in the short run; the disadvantage to reproduction would soon be overwhelming.

Few females in their breeding prime (such as Scratch's mother) could risk handicapping their own reproductive careers by helping another animal—even a close relative—in a dangerous fight. Self-sacrifice does occur, but it is exhibited by older female relatives approaching the end of their careers. For females such as old Sol, the reproductive cost of altruism in defense of her relatives would be very small, and the advantage would be real because their shared genes would be preserved by their act of defense. Younger females with more to lose must be more cautious.

Genetic competition between females also handicaps the resisters in another way. If infanticide really is advantageous behavior for males, and if it is (as I believe) an inherited tendency, any female who sexually boycotted infanticidal males would do so to the detriment of her own male progeny. Her sons would inevitably suffer in the ruthless competition with the sons of less discriminating mothers.

Returning to the question posed by Gilbert and Sullivan,

"If you enlist all women in your cause, and make them all abjure tyrannic Man . . . How is this posterity to be provided for?" the answer of course—as illustrated by Princess Ida's own handmaidens—is that somebody is going to cheat, and it is precisely her genes that appear in the next generation. No matter what cost consorting with infanticidal males imposes on her sex as a whole, no single female can afford to boycott a father with such an advantageous legacy for his sons. In the parlance of sociobiology, the altruistic female who refuses to mate with an infanticide is pursuing an "evolutionarily unstable strategy." Such a strategy is easily bested by a competing female. Over time, that strategy would be selected right out of the genetic repertoire of the population.

In case after case throughout the natural world, animals are caught in similar evolutionary traps: selection favoring individual gain detracts from the fitness of others, from the general viability or survival of the species, and from what humans might call "quality of life." Infanticide simply happens to be a particularly striking and well-documented example of this larger phenomenon.

Anyone who for even a moment thinks that what is natural is necessarily desirable has only to remember that 90 percent of all species that ever evolved are now extinct—through natural processes. Although natural selection may have had a role in shaping such values as love of children and revulsion toward those who harm them, the process of evolution itself is oblivious to such sentiments. From an evolutionary perspective, the chances that either males or females will break away from the system sufficiently to control their own destinies seem slim indeed. But the same perspective also lends an element of heroism and larger purpose to missions which aspire, despite the odds, to literally change the rules of existence.

INFANTICIDE by rival males provides a classic example of reproductive exploitation of one sex by another. In this instance, males compete with other males and pursue genetically selfish strategies at the expense of females and their offspring. Less often noticed, however, is the necessary underpinning for the evolution of this system: competition between females themselves.

As I hope to show in the next chapter, the two most salient forces at work in the social arrangements of primates are competition among individual females and cooperation among related females who are competing with other groups of related females. Whether primates live in pairs, like marmosets, in harems like some langurs, or in flexible matrilineal assemblages patrolled by brothers, as chimpanzees and spider monkeys do, the social system is dictated by how females space themselves and by the hierarchies they establish. The basic dynamics of the mating system depend not so much on male predilections as on the degree to which one female tolerates another. This tolerance is often little more than a matter of diet. This may seem like an odd perspective to take, but if male breeding strategies are determined by how females space themselves, and if that, in turn, is determined by the availability and utilization of resources, food becomes, quite literally, the consuming question. Of all the chapters in the book, I consider this next one the most important.

The failure of the women's movement past and present . . . can be attributed to three causes: the failure of women to bond; the failure of women to imagine women as autonomous; and the failure of even achieving women to resist, sooner or later, the protection to be obtained by entering the male mainstream.   CAROLYN HEILBRUN, 1979

6

Competition and Bonding among Females

The complexity and richness inherent in the social networks female primates forge for themselves has, too often, obscured a vital fact of their lives: that competition among females is central to primate social organization.

One's first glimpse of a troop of monkeys is far more likely to reveal a cluster of females picking intently beneath the fur of a relative to rid her skin of parasites than it is to catch a pair of females fighting. From day to day, companionable behaviors such as grooming or clustering together are far more frequent and visible than competition. If the troop is endangered, what stands out is often female solidarity. Sugiyama, Mohnot, and I have each watched langur females charge a male in order to rescue another female's offspring. Squirrel monkeys or savanna baboon females may mob an animal that endangers an infant—half a dozen screaming females throwing themselves upon an offender with unanimous indignation. Joint defense of feeding grounds by troopmates is commonplace. For many species, infant sharing, in which females take turns carrying newborn infants, is a routine part of child bearing.

But these allies are also the competitors with whom a female is in closest proximity, and they vie for the same resources. From species to species, the relationship between two females may grade from friendly to aggressive, but beneath the diversity there runs a common theme: every female is essentially a competitive, strategizing creature.

A venerable tradition in the behavioral sciences of focusing upon males is only part of the reason that competition among females has been so long ignored. There is a second problem: relations among females are, quite simply, not always what they seem on the surface. Whereas fights between males over females, or conflicts between males and females, take overt and even startling forms, persistent conflicts between females are often more subtle. Occasionally competition among females is expressed openly—especially when unrelated females from different troops meet—but more often competition is indirect. Two animals who are not even touching or looking at one another may nevertheless be in competition if one of them occupies a place or consumes a resource that would benefit the other.

Infant sharing, with all its complexities, provides a striking illustration of the fine line between cooperation and exploitation, between a female helping a relative in her troop and one helping herself at a relative's expense.

In virtually all monkeys, newborn infants exercise a magnetic attraction for other females in the troop. Shortly after birth, troopmates cluster around to inspect the infant. Some females will attempt to pull the neonate off the mother, hold, and carry it. Virtually all females are eager to get their hands on newborns, the only exceptions being some older, very experienced females. The main variable from species to species is the willingness or reluctance of the mother to give her infant up. Among squirrel monkeys, howlers, vervets, and virtually all colobines, most mothers readily give up their newborns. A newborn langur may spend up to 50 percent of the first day of life in the possession of "allomothers"—females other than the mother. By contrast, baboon and rhesus macaque mothers typically refuse to give up their infants for some weeks after birth. In bonnet macaques, only low-ranking mothers, who presumably have little choice in the matter, give up their new infants.[1]

Infant sharing has various advantages. The mother, for one,

may gain freedom to forage unencumbered by a clinging baby.[2] There is the prospect of additional protection—even adoption—of infants in case the mother dies or is temporarily incapacitated.[3] Perhaps most importantly, inexperienced young mothers-to-be have an opportunity to gain practice by carrying and caring for an infant. The hypothesis that allomothers are learning maternal skills gains support from the finding that for rhesus macaques, langurs, and some populations of vervet monkeys it is the nulliparous females—those who have never had an infant of their own—who are the most eager and assiduous caretakers.[4]

On the negative side, the experience of infant sharing can be trying and even hazardous for the newborn. Whereas some allomothers treat their borrowed infants with remarkable gentleness, occasionally exhibiting even greater solicitude than the mother herself, other allomothers may abuse or abandon their charges. Among the hierarchical cercopithecine species such as macaques and baboons—whose troops are composed of several different matrilineages and whose multimale breeding system means that animals in the same troop are on average less closely related than in a one-male breeding system—a high-ranking female who takes an infant may refuse to return it, with the result that the infant starves to death. Among the more closely related females in a langur troop, on the other hand, the risk of another female harming the infant is much less; mothers are virtually always able to retrieve their infants except in those rare instances when an infant is kidnapped by females from *another* troop, incidents which usually end in starvation of the infant.[5]

Nulliparous females, while the most eager of the allomothers, are occasionally also incompetent. Pregnant and lactating females may also borrow newborns, but such females tend to lose interest sooner. In the process of getting rid of their clinging passenger, allomothers may push it off with one foot, step on it, sit on it, or drag it along the ground. A screaming infant abandoned by an allomother quickly attracts the attention of either the mother or another caretaker, and abandoned infants are rarely on their own for longer than a few minutes. Nevertheless, the extreme vulnerability of infant primates clearly places such an infant at some risk.

For years it was assumed that allomothers were simply helpers in a communal child-rearing system. Indeed, allomothers probably are helping the mother to some extent. But the recurrent observations of abuse suggest that not all females who borrow infants are doing so merely to care for them. Most allomothers are also exploiting their charges in some way, using them as pawns in social interactions or as props to practice maternal skills.

There is probably no better example of just how subtle and how complex competition among females can be than the effects of one animal upon the menstrual cycle of another. In some cases, reproductive inhibition of subordinates can be brought about merely by the presence or seemingly unrelated activities of a dominant female. In other cases, there is overt harassment of the subordinate by the dominant. Among a variety of animals (including such primates as marmosets, tamarins, gelada baboons, savanna baboons, talapoin monkeys, and various macaques) the presence of dominant females may be implicated in delays in maturation, inhibition of ovulation, or, in extreme cases, spontaneous abortion by subordinates. But it took years for animal keepers and scientists to realize what was going on; the mechanisms of reproductive inhibition still remain obscure, although experimental work has begun to elucidate the relationships between subordination, social stress, and such important components of reproductive physiology as estrogenic hormones.[6]

More subtle, and even less clear, is the mechanism which causes mothers in some species to produce more sons than daughters. Detailed examination of birth records for captive and wild galagos, particularly *Galago crassicaudatus* (the "thick-tailed bushbaby"), reveal that mothers give birth to a remarkably higher proportion of sons than daughters. These nocturnal African prosimians make their living catching insects and eating the gum of acacia trees. Whereas maturing sons leave their mothers' feeding grounds to find a territory of their own, galago daughters linger on. Sons move on to compete with other animals, but daughters compete with their mothers for locally available resources. Primatologist Anne Clark has recently suggested an intriguing hypothesis to explain this uterine bias against daughters among galagos: mothers are avoiding competition with local females simply by producing fewer of them.[7]

Competition among females can be subtle, and also fraught with ambiguities—again, that fine line between cooperation and competition. Female primates do cooperate with one another. They do bond, but they do so imperfectly or incompletely. They cooperate selfishly, and there is a perpetual undercurrent of competition. The aim of this chapter will be to examine the forms this competition takes and to illustrate the impact it has upon the lives of females, particularly as it affects the status of females relative to males.

A MONG polygamous primates, groups of females (usually relatives) provide each other with indispensable support. Without this kin-based solidarity, females are abjectly subordinate to a male leader, as they are among gorillas and hamadryas baboons. Even with such a support system, males typically, though not always, remain dominant on a one-to-one basis. That female solidarity provides crucial leverage against males is scarcely a novel observation; it is one which occurred long ago (and without benefit of primatological field studies) to the feminists who coined the term "sisters" to apply to allies in a common power struggle. But if we confine our discussion to nonhuman primates, there is no more striking illustration of the importance of female solidarity than a comparison between two species with superficially rather similar social systems: the hamadryas baboon (*Papio hamadryas*) and the gelada (*Theropithecus gelada*).

Unique among nonhuman primates, hamadryas and gelada live in complex, multitiered societies. The basic social and foraging unit is a harem of females accompanied by one, sometimes several, adult males. One male is typically dominant and responsible for most of the breeding. These independent one-male units, containing from two to ten animals,

Hamadryas baboons

Gelada baboons

assemble into enormous herds which congregate nightly at crowded sleeping cliffs or come together seasonally to utilize widely dispersed food sources. Herds may number into the hundreds, but they are far from casual associations. Units recognize one another and distinguish herd members from outsiders. Relations between herds are antagonistic. Whereas females belonging to other herds are fair game, within each herd powerful inhibitions prevent males from so much as interacting with a female in another male's harem.[8]

Despite such similarities, from a female's point of view life in gelada and hamadryas herds are radically different. The female hamadryas is probably the most wretched and least independent of any nonhuman primate. Whereas the cowering hamadryas might not be envied even by one of Bluebeard's harem, the gelada female, like Lysistrata, is submissive only to a point; sufficiently provoked, she and her female allies take the offensive against males. The difference between the two species is rooted in the quite different company each female keeps. Beneath the superficial similarities there are profound differences in the ecological adaptations and social structure of the two species.

Isolated populations of geladas, hidden away in inaccessible gorges of the central Ethiopian highlands, are relics of seed-eating and grazing monkeys that once ranged throughout the vast plains of eastern and southern Africa. They are relics of a genus which has fairly recently known better days. By contrast, the hamadryas have never been more than a poor relation of the highly successful group known as savanna baboons. Few places on earth could be as forbidding as the arid homeland of the red-faced hamadryas, areas like the Danakil plateau of Ethiopia where even acacia thorn trees are so scarce and stunted that a relatively small-bodied hamadryas male sitting in one looks gigantic and out of proportion. At present, hamadryas baboons are found only in the badlands of the eastern Sudan, in the eastern lowlands of Ethiopia and Somalia, in Saudi Arabia, and in a small area of Yemen. The flexible hamadryas social organization—large herds that break up into smaller units for foraging and then reassemble at night about the few suitable sleeping cliffs for protection from leopards—is an adaptation to this harsh environment. Food resources (mostly seeds, occasionally fruits or shoots) are patchily distributed and unpredictable.

The hamadryas probably originated as an offshoot of savanna baboons living in richer habitats to the south. Physically, hamadryas resemble contemporary savanna baboons, but they are much smaller than their black-faced relations. In the males, this size difference is masked to some extent by a thick mantle of fur which frames a rather threatening face the color of raw beefsteak.

One reason the hamadryas survive at all is the males' unique mode of acquiring a harem—a system described by the Swiss ethologist Hans Kummer in his now classic study of the relationship between environment and social organization.[9] In the hamadryas' energy-scarce environment, fighting would be costly indeed. Instead of fighting with established males over their troop of mature females, young adult males in the process of building a harem "kidnap" or lure away juvenile hamadryas females, with a minimum of resistance from the parents. From the first day, the immature female is conditioned by her captor to unfaltering obedience. Wherever her male leads, she follows. If she hesitates, she is stared at (a threat), chased, and herded. If she balks, she is bitten on the scruff of her neck. A female who strays even a few meters for an unscheduled drink of water is chased and bitten—occasionally with such force that she is lifted right off the ground. In only very rare instances, however, has a field worker seen a female bitten so repeatedly by an overzealous consort that her life was endangered.[10]

Despite the seeming brutality of this system, the behavior of hamadryas males is probably best understood as excessively paternalistic. From the male's point of view, he has nurtured this female from preadolescence to breeding age. For several years, he has guided her each morning to harvest the flowers, fresh shoots, or fallen pods from acacia trees. When necessary, he has carried her up particularly steep cliffs. On rare occasions, when his possession of her was threatened by another male, he has fought for her, or else wrapped his arms about her and covered with his body this diminutive female who even in adulthood will be only half his size. His investment is substantial: he is protector as well as tyrant.

There is, of course, a female side to the hamadryas story, though only a fraction of it is actually known. From Kummer's field studies it is clear that herder and herded are co-adapted. After a period of conditioning, the chastised female quickly learns to follow. Her compliance toward males is virtually unique among monkeys and essential to the working of the system. Every so often, in the border zone between hamadryas and its savanna baboon neighbor, a male from one species will attempt to spirit away a female from another. When a hamadryas male attempts to herd a savanna female introduced from an adjacent population by experimenters, the more independent savanna baboon female simply runs away, leaving the perplexed male to vent his frustration in the dust. An anubis–hamadryas hybrid male (for the two species are still close enough to interbreed) living amidst hamadryas baboons is at a serious disadvantage. Innocent of the urge to herd and nip females, he is also unable to keep any.[11]

But the most remarkable fact about the hamadryas system is that when the male who has gathered the harem together is incapacitated or removed, the group disintegrates. Unlike all other social monkeys, and all but one of the apes (gorillas too have male-centered harems), female hamadryas baboons merely tolerate one another; they never bond. Hamadryas females without a male disperse, like so many pigeons from a cage.

Theirs is a very different society from that of the gelada, where females—most of them related by birth—are so tightly knit that relationships among females remain more or less unchanged by the comings and goings of males. There have been two detailed studies of these remote monkeys, one by the British primatologists Robin and Patsy Dunbar, the other by a team of Japanese scientists under the leadership of Masao Kawai.[12] Their studies took them high into the Semien mountains of Central Ethiopia.

Picture thick mist settling over an escarpment, drifting down from granite outcrops and enveloping spiky lobelia plants which dot the alpine meadows like giant tumbleweeds tethered to stalks. Hirsute brown forms shuffle along the meadow floor on padded bottoms, picking blades of grass. Pink blisters stand out like beadwork on the bare chest patches of some of the females; these curious configurations lend geladas their popular name as "the bleeding-heart baboons." Adult males are adorned by both bald patches on their chest and contrasting thick mantles of hair about their head and shoulders, resplendent as witchdoctors' headdresses. Quite possibly the esteem of natives for these manes, which substitute in some initiation ceremonies for the manes of lions, have, along with other pressures, brought this beleaguered genus to the doorway of extinction.

A disturbance at the periphery of the herd focuses attention on a female who has strayed from her unit. Her leader has moved to retrieve her. Herein lies the core difference between hamadryas and gelada society: the male rushes at her; the female snarls and lunges back; she is joined by three other females from the same harem who stand their ground beside her. Together they chase the male.

Whereas Kummer had reported for hamadryas baboons that "no evidence of bonds among adult females" could be found, both teams studying geladas stressed the strong mutual attraction among females in harem units. As a result of such solidarity, the position of gelada males is considerably weaker than was previously assumed on the basis of the general similarity between gelada and hamadryas herd structure. Whereas a hamadryas male gathers together females from different sources, making him the main cohesive force holding his eclectic harem together, a gelada male usurps or inherits an intact harem of female relatives. As with langurs, all-male bands haunt the vicinity of the harem and wait an opportunity for one of them to take over. Although the Dunbars reported some movement between units by juvenile females, the Japanese stressed the permanency of female affiliation. Like patas monkeys or langurs, females remain in their natal group and exhibit considerable solidarity in the face of other units or males who enter the troop from outside it. Typically, they ally themselves against these common antagonists.

The primary source of gelada solidarity is kinship, synonymous in this and many other primate societies with close association from birth onward. But there are limits to kin-based alliances—as we saw in the case of langurs. And there is also a dark side to the benefits females gain from banding together. Whereas the assertiveness and solidarity shown by females in such nepotistic species as macaques, baboons, and geladas is advantageous for high-ranking females, such systems lead to persecution, suppression, or exclusion of other females who are either low-ranking animals within the group or else outsiders belonging to a neighboring group.

Among the close-knit gelada, relationships between harem females are characterized by continuous low-level competition, much of it subtle jockeying for position near the adult male leader of the troop—the same male that under duress females unite together to chase! Both the Dunbars and the Japanese team noted a hierarchical ordering of females in each harem. The top or alpha female, or the top two females, enjoy a particularly close relationship with the dominant male. These high-ranking females routinely interfere with other females, particularly estrous females seeking proximity to the male. The alpha female may threaten any estrous female who approaches the leader, or may squeeze between the leader and the soliciting female to prevent them from touching or grooming. Rarely, though, does this jockeying erupt into outright aggression. Either through temperament or a lack of natural weaponry (the canine teeth of female primates are almost invariably smaller than males' except in the case of some monogamous species), females rarely inflict serious damage on one another in their quarrels. Even though female–female quarrels are more frequent than fights between males in most species, encounters between adult males are much more likely to result in one or both animals' being wounded.

In the short term, the advantage of dominance to a female is that she is first to gain access to physical resources. When several gelada units arrive simultaneously at the same waterhole, the male and dominant females drink first and perhaps wallow in the water; subordinate animals wait their turn. Once the unit leader has drunk his fill and moved on, other units may sometimes pressure subordinate animals to move from the water before they have had a chance to drink. Subordinate females fare no better when preferred foods are at issue. A gelada who wants to get at the prized heart of a lobelia plant must first break away its spiky outer leaves. Males, who are stronger, are better able to expose the soft inner parts of the plant; dominant females closest to them also share in this privileged access. Even if a subordinate female does find a feeding site of her own, she may be bullied into giving up her position to a more dominant female.[13]

Several studies suggest that there may be long-term consequences from this low-key jockeying for position. When the Dunbars determined dominance rank for all the females in each of 11 gelada units (by recording all interactions in which one animal approached and the other withdrew) and then also counted the number of offspring belonging to each female, they found a significant correlation between a female's rank and her reproductive success. They hypothesized that this small but statistically significant differential between females was due to harassment of subordinates and that such harassment is sufficiently stressful to reduce fertility.[14]

It was by a process of elimination rather than direct evidence that the Dunbars arrived at their conclusion. From field observations the Dunbars knew that both high- and low-ranking females had had ample opportunities to copulate. Hence, if low-ranking females were not becoming pregnant, it must be due to some internal process. The Dunbars knew from experimental studies of captive savanna baboons, as well as studies of a number of smaller mammals such as rodents, that intense aggression or stress can alter endocrine profiles, attenuate menstrual cycles, reduce the period of sexual receptivity, or actually cause reproductive failure. In rats, reproduction can be so sensitive to physical stressors that individuals whose mothers were stressed while they were still *in utero* showed the effects at maturity. Daughters born to these stressed mothers had significantly reduced fertility.[15] Because of the long generation time in higher primates, the expense of keeping them in captivity, and ethical considerations, the relationship between stress and fertility is much better understood for rodents than for members of our own order. For no primate do we have data on lifetime reproductive success under natural conditions, much less experimental data contrasting reproductive careers of stressed and nonstressed females. However, what evidence there is clearly suggests that for primates, as for rodents, stress is a factor in reproduction.

Several observers in addition to the Dunbars have noted that for whatever reason, high-ranking females in a given cage or troop are typically first to get pregnant. Thelma Rowell, studying six different species of Old World monkeys kept in her breeding colonies, reported that "in all cases the highest ranking females were the first to breed, and, with few exceptions, females became pregnant in order of descending rank." High-ranking females also tend to wean their infants sooner. In both captive and wild baboon populations, the menstrual cycles of females beaten up by other females were lengthened, so that they came into estrus less often.[16] Furthermore, where hormonal information is available for primates, endocrinological profiles for subordinate animals are characteristic of those of "stressed" animals. Subordinate female talapoin monkeys, for example, exhibit high cortisol and prolactin levels, and subordinate females fail to respond to injection of estrogen with the surge in luteinizing hormones essential for ovulation.[17]

But why should females respond to stress in this seemingly nonadaptive way? Although the gelada case is less extreme than that of subordinate marmosets, who cease ovulating altogether, it is clear that any female who, over time, reproduces at a slower rate than other members of the population puts herself at a genetic disadvantage. So why would a female evolve the capacity to respond to low status in this way, by reducing her rate of reproduction? Various explanations have been suggested. Possibly, the stressful conditions of second-class citizenship simulate unfavorable environmental conditions such as food shortage. Deferment of reproduction for a malnourished mother or stressed animal would be adaptive insofar as it allowed her to conserve resources until a more favorable opportunity to breed arises. By ceasing to reproduce, the female avoids dangerous depletion of her bodily resources during a difficult period.[18] This explanation assumes that a low-ranking female would have opportunities some day to improve her lot, to turn the tables, become dominant in her own right, or leave the group. Other possible explanations are that low-ranking females are actually deprived of nutrients crucial to reproduction, or that it takes a subordinate animal longer to stockpile sufficient fat and other reserves to carry her through a metabolically expensive pregnancy and lactation.

Carried to an extreme, harassment of low-ranking females may lead to murder of her offspring, but such occurrences must be very rare. Females are all about the same size, and an infanticidal female would not have as great an advantage over the mother as a male does. To date, infanticide by female associates of a mother has been reported only for wild chimpanzees and gorillas, and only in the chimpanzee case was the incident actually witnessed. On three occasions at the Gombe Stream Reserve, Jane Goodall's field assistants witnessed females belonging to a particular high-ranking lineage ("Passion's") murder the offspring of other females who in each instance were both lower ranking than the infanticidal females and were also hampered in their capacity to retaliate by some physical disability. Gilka, for example, the low-ranking mother whose offspring were murdered by Passion or her daughter two years running, suffered from partial paralysis of her wrist and hand and from disfiguring facial fungus. The mother of the other victim suffered from a partially paralyzed neck since she, like Gilka, had once had polio.[19] Chimpanzees and gorillas, along with hamadryas baboons, red colobus, and South American howler monkeys (and some human societies), are the only primate groups where females are known to routinely migrate between groups or communities; therefore, except for particular mother–daughter clusters, females are not on average close relatives. Hence, one of the checks on intragroup aggression normally present among female primates—close genetic relatedness—is greatly reduced in the chimpanzee and gorilla cases. Female–female cooperation, particularly communal infant sharing, are uncommon; at most, an older sibling will be permitted to handle her mother's new offspring. Still, the motives promoting infanticide by female chimpanzees remain unclear. Was it purely predation? Each victim was eaten by the female who killed it. Or were females from the dominant lineage eliminating a competitor for resources while the infant was still sufficiently vulnerable to be dispatched with impunity?

Regardless of what the exact motives or mechanisms are, it is clear that high rank carries with it not only freedom from harassment and exploitation by more dominant females but also this sinister prerogative to interfere in the reproduction of other females. Generally speaking, the day-to-day advantages of rank are quite small, typically involving food resources and the ability of one female to displace another. Often it is just a matter of avoidance; the subordinate animal simply steers clear of interactions. On a daily basis, the connection with reproduction is remote. But over time, and especially generations, the small effects of female–female competition become enormous. A host of questions comes to mind. What perpetuates these dominance systems? Why do low-ranking animals comply—or do they? The best documented answers to such questions derive from long-term observations of those sturdy little brown monkeys, the macaques, and from several studies of savanna baboons.

THERE exists no better illustration of the profound influence of inherited rank in the lives of animals than the genus *Macaca*. For many years scientists in Japan and on several small islands in the Caribbean have monitored free-ranging colonies of *Macaca fuscata* and *M. mulatta*. *Macaca fuscata* is the macaque native to Japan, where for generations the Shinto inhabitants have regarded these monkeys with the same mixture of reverence and annoyance that Hindus reserve for langurs. To protect them for study, and also, as it happens, to use monkeys as a tourist attraction to bolster local economies, they were lured down from their forested hillsides to beaches and cleared areas surrounding feeding stations. Before World War II a boatload of the related species *Macaca mulatta* was imported from India to Cayo Santiago and La Parguera Islands off the coast of Puerto Rico and released. For decades, births and deaths among these free-ranging macaques were recorded; the resulting longitudinal data provide ages and genealogical relations for numbers of animals.

With remarkable regularity, these macaques arranged themselves into large multimale troops composed of ranked matrilineal clans. Adjacent troops were ranked; within troops, matrilineal clans were ranked; and within the clans, individuals. At the top of each matrilineal hierarchy reigned the founding female. Although dominant males in a troop may outrank the top-ranking matriarch of each clan, the rank of such males still depends on female support. By comparison with the high-ranking females of the top matrilineal clans, male power is transient.

Females remain in the same troop for a lifetime, whereas males transfer out after a few years. Among Japanese macaques, males leave as early as one or two years of age. Out of 152 *Macaca fuscata* males marked at birth, all had left their natal troops by the age of twelve.[20] Many of them spend time in roving all-male bands, called *hanarazaru* in Japanese, before re-entering troop life through a relatively peaceful process of assimilation. Males rarely remain in a troop more than five years before shifting again, possibly an evolutionary mechanism to reduce inbreeding, especially perhaps to keep males from breeding with their daughters. Despite constant flux in male membership, the rank of one matrilineage relative to another remains remarkably stable, suggesting that the personality of particular males has little to do with mainstream social processes. Rather, the social structure is determined by the energy and assertiveness of particular females.

Within each matriline, relationships between individual females are so stable that the outcome of an interaction can be predicted with all the regularity of a seating arrangement at a diplomatic dinner party. That is, there is a very strict and regular protocol marked by occasional tension-generating upsets. Females are ranked according to two rules: inheritance of rank from their mothers, and "younger-sister ascendancy." According to the first rule, daughters fit into the hierarchy just under their mother. According to the second, a younger daughter at just about the time of her first pregnancy rises in rank above her elder sister.[21] Several field observers have noted that the rank reversal between sisters is, at least in part, brought about by the mother, who tends to intervene on behalf of the younger daughter.

One recent explanation for the curious phenomenon of younger-sister ascendancy is that insofar as the matriline as a whole is concerned, the younger female is more reproductively valuable since she still has the peak years of reproduction before her.[22] Another hypothesis, not mutually exclusive with the one just mentioned, stresses the fact that first pregnancy is a critical phase in a macaque's reproductive career. The young female, pregnant for the first time, is particularly vulnerable and so is her infant. For populations of free-ranging rhesus macaques where detailed demographic data are available, as many as 50 percent of firstborn infants die in the first six months, compared to much lower mortality rates for later born infants. Similarly, among squirrel monkeys monitored in a laboratory setting and wild howler monkeys, 70 percent or higher of firstborn infants die. According to this second hypothesis, then, the mother intervenes on behalf of the younger daughter, helping her to rise in rank above her older sister, to enable the younger daughter to reproduce sooner (the timing of first pregnancy tends to be rank-related) and perhaps especially in order to improve the survival statistics of these highly vulnerable firstborn infants.[23] Whatever the explanation, the outcome of maternal determination of rank is a highly nepotistic society where individuals inherit unequal lifetime benefits according to the happenstances of birth. Preliminary data for other species belonging to the cercopithecine branch of the Old World monkeys, including other species of macaques from Malaysia and Borneo, vervet monkeys, and several species of savanna baboons, have led primatologists to speculate that nepotism may be the general rule for cercopithecine monkeys living in large multimale troops composed of several matrilines.

Among such species, individual status and the status of lineages is a serious matter, aggressively defended. Small wonder, since over generations the stakes become enormous. Demographic data for troops of rhesus collected over a ten-year period on La Parguera Island indicate that: (1) a larger percentage of high-ranking than low-ranking females breed each year; (2) offspring of high-ranking females have a higher rate of survival than do infants of low-ranking females; and (3) the daughters of high-ranking females themselves produce offspring at a significantly early age (3.85 years) compared to daughters of low-ranking females (4.4 years).[24]

So predictable are these nepotistic systems that Glenn Hausfater, a scientist not known for exaggeration, claimed in a recent address at a scientific meeting that if he were given the age, sex, and maternal rank of an individual, he could tell you three-quarters of what one needed to know about that animal. He was referring specifically to the baboons of Amboseli (*Papio cynocephalus*). Together with Jeanne and Stuart Altmann, Hausfater has monitored life histories of a troop of Amboseli baboons ("Alto's troop," named for a now-deceased matriarch) since 1971. They have learned that the lives of those animals can be predicted with considerable accuracy simply by knowing, for example, who the mother was. Not only behavior patterns but other features of baboon daily life such as diet, the budgeting of time spent foraging and resting, or even the average number of parasite ova emitted in a stool—all of these tend to vary with the rank and reproductive status of the animal.

But how are such systems brought about? Knowing that dominant and subordinate monkeys have different endocrinological profiles or diets might explain why they have different susceptibility to disease, or why they reproduce on different schedules, but it does not tell us why the newborns enter the hierarchy in different positions in the first place. Hausfater suggests that for each infant at birth, the world can be divided in two: between those females who may approach the mother with impunity and attempt to pull the infant off, and those females who would not dare to! So fixed would be the resulting system that even if the mother dies prior to maturity, the daughter still fits into the hierarchy at the same spot the mother did, or just above her older sister, if she has one.[25]

It is not yet known what effects maternal rank has on the reproductive and life histories of sons, but preliminary data for both macaques and baboons suggest that the sons of high-ranking mothers may enter the hierarchy with an initial advantage. If so, this might mean that sons of high-ranking mothers would be more likely to remain in their natal troop, or at least more likely to remain there longer. One need only assume that at puberty all males attempt to rise in rank. Sons of low-ranking mothers would probably have to leave the troop in order to do so, but sons of high-ranking mothers might have an opportunity of achieving high rank within their natal troop, and hence linger on.[26] Whatever advantages the mother's high rank might mean for her sons, they are probably less important than the advantages daughters enjoy. One fascinating consequence of this system is that high-ranking females in nepotistic species such as savanna baboons and bonnet macaques tend to produce more daughters than sons. In contrast, low-ranking females produce fewer daughters and more sons. The explanation seems to be that since males, in any event, are liable to leave their natal troop, a low-ranking mother's status is less of a liability for sons than it would be for daughters and she therefore produces more sons.[27]

The precise mechanisms that lead to earlier maturity,

higher infant suvivorship, and generally greater reproductive success for high-ranking rhesus females from La Parguera are not known, but studies of two other macaque species, the toque macaque and the Japanese macaque, suggest that competition for food may be one important variable.[28] For over a decade Wolfgang Dittus has studied the demographic consequences of social competition among a population of 500 toque macaques (*Macaca sinica*) on their island homeland in Sri Lanka. Like many macaque species, toques live in troops with one or more adult males. In common with other macaques, toques live off fruits, seeds, and gum. When such foods are very abundant, handfuls are crammed into expandable cheek pouches for later mastication. These pouches are so well engineered for competitive gluttony that they permit preliminary digestion of starches to begin while stuffing is still in progress. Toques are easily distinguished from other macaques, however, by their arboreal habits, their tousled equivalent of a mohawk hairdo, and by small body size—an adult male weighs 6 kilos (about 13 pounds), the female just under 4 kilos (9 pounds).

Competition for food is extremely intense in the lives of these monkeys. Eighty percent of all toque-to-toque threats occur during foraging. Dittus notes that a threat during foraging may prevent one animal from approaching another. Or it may cause a threatened animal to sit still and stop feeding while a dominant animal feeds nearby. More usually (36 percent of the time) the subordinate is simply displaced and forced to move away, while the dominant animal feeds in its place.

Usurpation of food by dominants is sometimes carried to extremes. An adult male who had recently fallen in rank and had been forced to the periphery of the troop was observed nonchalantly removing and consuming the contents from the cheek pouches of an older juvenile female. According to Dittus, this incident occurred during a period of general food shortage when the consequences for the subordinate animal would be severely felt.

When food was scarce, rank differences were reflected in

Toque macaques

lower foraging efficiency and lower body weights for subordinate animals who were displaced from the richest feeding areas. During the first six and one-half years of Dittus' study, the toque macaque population fluctuated from year to year but on the whole remained constant in size. Significantly, periods of food shortage coincided with higher rates of mortality among subordinate animals—usually juvenile. Dittus concluded that it was competition for food which was keeping the population within the observed bounds.

Similar data supporting Dittus' hypothesis that population size is regulated by socially mediated competition for food have been collected by Akio Mori in Japan. Mori took advantage of unusual circumstances surrounding the maintenance of a single troop of 20 Japanese macaques at Koshima Islet. All macaques in the area had been designated a "national treasure" by the government of Japan. For some years (between 1952 and 1963) this semiwild troop had been sporadically fed by local people. But beginning in 1964 food was provided *ad libitum*. This provisioned population burgeoned. The density at Koshima Islet leapt up to 343 macaques per square kilometer, roughly tenfold the density of a wild population in a similar habitat. Then abruptly, in 1972, provisioning ceased. Akio Mori was on hand to record the consequences: a sharp decrease in birth rate; a higher death rate for older females, juvenile females, and especially infants; and a postponement among maturing females of the age of first reproduction. Whereas previously females had reached sexual maturity at age five, as food became scarce females did not engage in sexual behavior before the age of six. Weights and measurements were taken both before and after the period of food shortage. After provisioning stopped, all females weighed less, but the adult daughters of high-ranking mothers tended to maintain their weight better than did low-ranking females, and infant mortality among these smaller mothers was higher. From data collected over the next five years, Mori noted that "both the mother's and baby's survival depend on the mother's ability to maintain her body weight after parturition." The primary options open to a female were to rely on matrilineal relatives for protection, or to stay near to one or more of the troop's dominant animals who could enter the center of the troop and feed where he chose.

Survival and the production of healthy offspring is not, then, just a matter of head-on competition. In few other mammals do affiliative relations and alliances, together with shared responsibilities in the rearing of infants, play a greater role than they do in primates. A host of behaviors—huddling on a frosty night, mutual grooming, tension-reducing hugs, and general tolerance and helpfulness toward troop offspring—are vital elements in the fabric of social life in monkey troops. But it is not possible to consider such cooperation separately from competition. It is a fact of macaque life that lineages sink or swim together, and this mutual dependence of females on their relatives explains what might otherwise seem a puzzling and curiously humanistic feature of macaque society: the high social status of older females which has been repeatedly documented for rhesus and Japanese macaques. Such a "matriarch" will continue to be deferred to by her younger, more vigorous, and often heavier or stronger descendants.

Matrilineal inheritance of rank and highly stable female hierarchies typify cercopithecine monkeys and possibly some of the apes (such as chimpanzees), but such systems are by no means universal among primates. Among the mantled howler monkeys of Central and South America and Hanuman langurs belonging to the colobine branch of Old World monkeys, rank fluctuates in the course of each female's lifetime. There is a general tendency for assertive young females to rise up through the hierarchy, reaching the top about the time they bear their first or second offspring, and then gradually fall with increasing age. A spot check of a howler or a langur troop at a given point in time is likely to reveal a social system in which young-to-middle-aged females occupy the top half of the female hierarchy, while very old and immature animals are at the bottom.

Howler monkeys are among the few species of primates in which both males and females leave their natal group to join new ones. In a preliminary analysis of howler social organization, Clara Jones has presented a provocative hypothesis concerning the ranking system of howler females. According to Jones, the quantity of palatable, nonpoisonous leaves that howlers feed upon is strictly limited, and there is competition among females for membership in groups which occupy suitable feeding grounds. Long-term survival and successful breeding will depend on whether or not a female is able to establish herself in one of these groups. Young animals either rise in rank or disperse. If females fail to achieve a high position in the hierarchy while young, they fail to succeed in the group as a whole. As Jones puts it, "The rule of dominance relations in mantled howler monkeys appears to be 'up or out.' "[29] Hence females are competing not for rank per se but for the opportunity to enter, or remain, in a group. One consequence of this system is that young females trying to enter a troop from outside have relatively little to lose, much to gain; they are liable to be more assertive and aggressive than are established group members who on average have less to lose.

The langur case is quite different. So far as females are concerned, langur troops are closed social units, and there is a tendency for troop infants born at the same time to be the offspring of one male. Offspring of the same age will usually be related to one another through both their maternal and paternal lines, as opposed to the macaque situation where troops tend to be larger and where several males breed simultaneously. It has been suggested that in the langur case, high degrees of relatedness among females in the same troop make it advantageous for older females to defer to their younger female relatives. From the perspective of each female's inclusive fitness—that is, the sum of her individual reproductive success plus the effects that her behavior has on the fitness of close relatives who share many of her genes—females approaching the end of their reproductive career should opt out of competition, thereby allowing more productive females first access to resources, and instead invest energy in the well-being of relatives. Such a model might explain the behavior of the very low-ranking old female Sol, who nevertheless took the most active and aggressive role in defending from infanticidal males infants born in her troop (Chapter 5). The langur model would benefit mutually both dominant and subordinate females, and it is difficult in their case to distinguish dominance relations from cooperation.[30]

THE nepotistic system of the cercopithecines is not, then, the only alternative open to primates. Given the disadvantages of low rank in nepotistic species, one cannot help but wonder why the system persists. Given the high stakes, why do subordinate females continue to participate in a system which discriminates against them? On an immediate level, compliance is a matter of necessity. Typically, dominance is backed up by powerful family alliances. Under duress, however—an increasingly constraining expansion of the dominant lineage, for example—subordinate lineages do sometimes fission off and leave the main body of the troop. Such splinters may concentrate in one (often inferior) portion of their original range, or else push into new territory. Among rhesus monkeys and olive savanna baboons, where such fissioning processes have been most carefully documented, splinter groups were led away by the highest ranking animal of a low-ranking lineage, typically an old female.[31] Most such splinters fail to establish any foothold in their new ranges, but a few flourish and expand, and may even lead (in rare instances) to the origin of new species.

Alternatively, when the opportunity arises, some animals may rebel. Macaque and baboon social relationships are remarkably predictable, but lapses in protocol do occur. Occasionally daughters do rise in rank above their mothers, or older sisters persist in bullying younger sisters long after both have reached adulthood.[32] Circumstances leading to breaches in protocol are not understood, but they can lead to fierce fighting and even murder. In many ways such lapses in protocol are more revealing than the humdrum working of the system: they illuminate underlying tensions and forcefully demonstrate that subordinate animals are not voluntarily inferior but suppressed. Such episodes also underscore the importances of alliances among close relatives. When rebelling females successfully advance the status of their lineage, almost always it is with the cooperation of lineage mates or by forging alliances with males outside the lineage. Rank upsets have been witnessed now among a variety of macaques, including caged and free-ranging rhesus, Japanese macaques, caged pigtails (*M. nemestrina*), and crab-eating macaques (*M. fascicularis*), as well as among savanna baboons in both forest and the more typical plains habitats. Such rebellions may be precipitated either by internal events, such as the incapacitation or death of an influential animal, or through external causes, such as the arrival of a new male. Male support may be a critical factor in the maintenance of high rank by a particular lineage or in the bid for higher rank by a subordinate one, just as female support is often critical to a bid for higher rank by an up-and-coming male. Once the usual stability of the troop is upset, fights to the death are not uncommon.[33]

The following account of a rebellion among crab-eating macaques in an experimental setting parallels observations from wild animals except that specific details about genealogies and the reproductive states of the females involved are more precisely known, and in this instance the precipitating circumstance was prolongation of a subordinate female's sexual activity by human artifice. Over a five-year period a colony of *Macaca fascicularis* kept in Birmingham, England, had been

socially stable, characterized by the clear preeminence of the
A matrilineage over the B lineage. Throughout this period a
close relationship existed between the dominant male Percy
and the foundress of lineage A. However, this social order was
completely overturned when the founding female of B lineage,
Betty, successfully rebelled. The reporting scientists attributed Betty's success to insertion of an intrauterine device
which rendered her sexually receptive month after month.
Thus equipped, Betty could subvert the special relationship
between Percy and members of the A lineage by entering into
prolonged consortships with Percy. While the foundress of A
lineage was pregnant, Betty and her recently matured daughter solicited Percy's support in interlineage conflicts. In the
course of the fighting, the matriarch of A lineage was severely
beaten. Her subsequent death permitted B lineage to rise
above A. Autopsy of the dead female revealed that her fetus
had been dead for some time, and the investigators suspected
a relationship between the fetal death and stress on the pregnant female.[34] Interestingly, a similar rank upset reported for
wild baboons at the Gombe Stream Reserve in Tanzania involved the overthrow of a pregnant alpha female by three subordinate females, including a mother–daughter pair. In this
instance also the toppled alpha suffered a miscarriage—apparently as a result of the usurpers' battering—but in this
case the mother herself survived.[35]

These incidents illustrate the role of female sexuality in
soliciting male support, a theme that will be elaborated in
Chapter 7. Nevertheless, it is important to keep in mind that
reproductive state is only one of several factors which induce
males to assist females.

In a detailed study of a multimale troop of South African
baboons (*Papio ursinus*), Robert Seyfarth was able to show
that males preferentially aided specific females, and not just
in return for sexual favors. In the majority of cases where
males intervened in an aggressive interaction on behalf of a
female, she was not currently in estrus. Only 17 percent of
such cases involved a male defending a sexual consort from
another female.[36] Seyfarth's most important finding was that
social interactions between males and females in his study

troop persisted throughout all reproductive states, including pregnancy and lactation, at rates comparable to those seen during sexual cycling. Put another way, male baboons may be "seduced" into supporting females, but immediate sexual favors are not the only grounds for male–female alliances.

For some years—perhaps partially by way of reaction against anthropocentrism in popular accounts—the flexibility and intelligence of higher primates has been seriously under-represented by the conservative phraseology of scientific papers, which have avoided attributing even a hint of con-sciousness to animals. Yet, recent analyses of baboon relation-ships, such as Seyfarth's, stress the opportunistic aspect of alliances. Implicit in these descriptions is a conception of strategy. A wide range of responses to any given social situa-tion is possible, and the concept of "instinct" as a fixed trait is not very helpful in considering them. Hence, it now seems probable that a baboon male does somehow take into account a range of contingencies—say, a long future of possible relationships with a female—when he defends her or her off-spring in a particular squabble. Furthermore, there is no evi-dence to indicate that males and females are qualitatively dif-ferent in this respect. Although I have stressed kinship as the basis for female alliances, there are cases on record not only of alliances between male and female but between females which seemed to be based on opportunism and mutual benefit rather than blood relationship. Hence, Seyfarth's co-worker Dorothy Cheney has documented two types of alliance forma-tion among female savanna baboons: alliances between rela-tives of adjacent ranks and alliances cultivated by low-ranking animals with high-ranking ones. "Social climbing" in Cheney's study was based on daily social contacts, casual grooming, and aid proffered in disputes between the higher ranking female and other animals.[37] More dramatic align-ments between animals in different lineages may be fashioned during times of social upheaval, as was described recently for savanna baboons living at Gilgil in western Kenya. Over sev-eral years, two primatologists, Barbara Smuts and Nancy Ni-colson, charted the dominance relations among these animals and established the existence of the expected hierarchy: a stable ranking of females who seemed to be following all the usual rules. Rather suddenly, however, this ranking was overturned after an important female ally of several high-ranking females accidentally drowned. Members of several high-ranking lineages, aided by a number of independent agents from middle- and low-ranking lineages, overthrew the four highest-ranking females. These four fell to the bottom of the hierarchy and were subjected to frequent harassment by animals who had risen above them.[38] Smuts and Nicolson had witnessed an exception to the known rules, a very important exception.

Now the perspective shifts. No longer are we bystanders walking beside a troop of monkeys, watching specific individuals; these are but fragments in a larger picture. Let us consider an entire species, and contrast this species with other species. Why do some females (such as rhesus macaques) congregate in large troops containing multiple lineages, hierarchically arranged, while others (such as Hanuman langurs) cluster in small harems composed of close female relatives? Why do still others (such as indri) space themselves singly, accompanied only by another adult male in a monogamous pairing, or else travel alone with their offspring, entertaining a male only occasionally, for the purpose of breeding (as do orangutans)? These questions are among the outstanding problems in contemporary primatology and are not yet wholly answerable.

Among the more ambitious attempts to answer them has been the work of the British primatologist Richard Wrangham, the most recent in a venerable roster of those who have tried.[39]

Several biologists have sought to put female primates, and the complex issue of competition between them, into biological perspective. (The Dunbars' emphasis on stress, for example, was essentially a socioecological explanation for the greater reproductive success of high-ranking females in the particular case of gelada baboons.) Wrangham proposes a general model to explain the social systems of primates. He starts with the assumption that competition for resources is the driving force behind group living. The consuming question becomes, quite literally, food, and competition for it. According to the model, females distribute themselves in space according to their needs and preferences for particular food resources, and their ability to utilize them. Species differ in digestive capacities. Whereas some species (like chimpanzees) are almost entirely fruit eaters, and hence must depend on intermittently available, high-quality foods which are clumped in space—a fruiting tree here and there—other species (like gorillas) are able to live off evenly distributed, more or less perpetually available low-quality forage such as leaves. These are the two extremes. The majority of monkey species are located somewhere in between: they combine subsistence on leaves with a preference for more easily digested and nutritious items like fruits when such are available. This mixed subsistence strategy has very special consequences for female bonding.

Once established on their feeding grounds, females either tolerate or exclude competing females. Tolerance of one female for another depends on how she makes her living. A primarily fruit-eating animal, such as a chimp, gibbon, or orangutan, depending as she does upon highly concentrated and transitory resources, can ill afford to share her larder with females whose needs are identical. Fruits ripen more or less simultaneously, usually in the same grove or circumscribed location. A number of frugivores clustered in the same location would deplete this finite resource before the needs of any one of them were satisfied. By contrast, leaf-eating animals such as gorillas look out upon a green world of more or less continuously available, evenly distributed, low-quality fibrous foods. Such foods, although not particularly digestible, are very abundant. While the frugivores must spread themselves thinly across their feeding grounds, more folivorous creatures can afford the luxury of clustering together in social groups.

But a folivorous existence is not quite as easy as it sounds. Plants have to make a living as well. To discourage animals from dining on them, plants manufacture a broad range of toxins and bitter tastes. There are strict physiological limits to the quantities of such plants that any one animal can process before she overloads her liver and is poisoned. This is yet another reason why leaf eaters can afford to share their slightly poisonous larder with competitors. Each animal uses only a small portion of each of the various species of plant available. For example, langurs possess exceptionally large stomachs divided into several compartments containing a wide range of anaerobic bacteria which help to break down plant toxins and tough fibers, thus liberating the sugars present for digestion. This ruminant-like system permits langurs to feed on such plants as *Strychnos noxvomica,* natural source of the poison strychnine, a delicacy that would kill many other nibblers. Competition for large quantities of this resource is unlikely; with unpalatable fare on the table, there is plenty to go around.

Reduced competition for food may explain why some species can afford to cluster together in large groups, but it does not explain why it should be particularly close relatives who bond together. Even leaf eaters, who can survive on roughage, compete for choicer items. New shoots and young leaves contain fewer toxins than mature ones. If females are competing for resources, would not a female do her relatives, and herself, a favor by moving away? In short, why don't relatives disperse?

There have to be advantages to remaining near relatives which offset this obvious disadvantage. Chief among these is defense of the family's feeding grounds against other matrilineages. Only by cooperating among themselves to defend their communal larder would relatives be able to maintain access to good feeding sites. Group cooperation would be necessary whenever food sources are distributed about the habitat so as to make them defensible—that is, in discrete patches, as is the case with groves of fruiting trees. A strong group could clearly monopolize such a resource by keeping other groups away. Competition within groups would simply be a disadvantageous side effect of this larger gain.

According to Wrangham, then, two things must be true in order to make it adaptive for females to remain near their relatives. First, monkeys must be able to subsist on widely available resources in the habitat, thereby reducing the cost to each individual of competing for local resources. At the same time, occasional utilization of defensible, high-quality foods (like fruits) makes it advantageous to live among cooperating relatives who hold their own against competing groups.

Wrangham's model applies beautifully to the Hanuman langurs that I know best. Langurs have the capacity to subsist on mature leaves, and in poor times they do so. But they prefer fruit, seeds, and new shoots. Langurs live in circumscribed home ranges which contain their feeding trees, but they sometimes trespass into trees growing in the range of a neighboring group if a preferred food happens to be available there. Fierce territorial encounters between troops may take place in areas where fruiting trees are close to the boundary between adjacent troops. Both males and females participate in the chasing, grappling, and slapping that ensues during such encounters. Because males whoop loudly and leap about boisterously, their participation is more conspicuous, and tends to mask the fact that the most persistent defenders of the troop's territories are often female relatives who inherited this feeding area from their mothers and grandmothers.

Closely knit matrilines who are intensely territorial fall at one extreme of a continuum of group-living primates. At the

Hanuman langur females in inter-troop encounter

other extreme are groups of females whose chief food sources are indefensible, who are nonterritorial, and who also are among the few species of group-dwelling primates without particularly close attachments between relatives. Females among gorillas, hamadryas baboons, and possibly also red colobus neither defend territories nor bond closely with female groupmates. These are species where females as well as males move between groups. In line with Wrangham's model, hamadryas baboons live in areas so arid that the animals must traverse vast areas to locate widely scattered food sources. High-quality patches of food, which might otherwise be defensible, constitute only a small portion of the hamadryas diet. Gorillas likewise find fruit on fewer than 2 percent of the plants that they use. Hence, a gorilla female loses little from having other females clustered about the male she has chosen, but neither does she benefit from the support of relatives in defending resources from other groups.

As in all reductionist efforts, Wrangham's model gains in generality by ignoring the intricacies of any real-life situation Complex areas of primate social life are scarcely considered; hence (to take one example) Wrangham pays scant attention to the role other groupmates play in rearing infants. In the years to come, the model will be argued over, refined, and possibly one day replaced. But for now, the basic outlines of primate social structure are better explained by Wrangham's approach than they have been by any previous model that I am aware of. Females arrange themselves in space and time so as to maximize food intake while minimizing competition for food from either individual females or from other groups of females. Within the limits compatible with survival, males arrange themselves singly or in bands so as to control access to these dispersed females. Males may shift strategies with age, physical condition, and opportunities, but the basic constraint upon them remains the deployment of females. Where females clump together (as among langurs or guenons) fierce contests between males over access to these harems develop. Where females are solitary, a male will either settle in with a single mate and help defend their common territory (gibbons, indri, marmosets, and titi monkeys) or move singly across the ranges of several mother–offspring units (galagos and orangs). The problems are far from simple and a host of questions remains. But the new focus on females has meant progress. The outcome has also led to a redefinition of the "nature" of female primates.

THE vision of assertive, dominance-oriented females differs radically from existing stereotypes of female primates as nonstop mothers whose perennial preoccupation with nurturing offspring keeps them out of politics. According to the old stereotype, "the number of adult males and their reciprocal relationships determine the social structure of the group as well as the group behavior as a whole."[40] As recently as 1978, in the first volume of its kind, entitled *Female Hierarchies,* neither the editors (Lionel Tiger and Heather Fowler) nor contributors to the volume could unearth much evidence that female primates were either competitive or that they routinely formed hierarchies. In her contribution on female hierarchies in evolutionary perspective, Virginia Abernethy wrote that among female primates, hierarchies are "difficult to identify and unstable." From the evidence she had at hand, she was forced to conclude that "little advantage from fighting for rank accrues to the individual because her preeminence will not last . . . it gives way before . . . protection of . . . the newborn." Hence,

> over evolutionary time there has been little functional advantage associated with the formation of female hierarchies. Insofar as the female usually has assumed responsibility for childbearing, and noting that there is little advantage to hierarchical organization when this is the task, a logical conclusion is that hierarchical behavior has never been a focus of selective pressure and that genetic predisposition toward the creation of hierarchy is not a major determinant of female behavior.[41]

In the last chapter of the same volume, Joseph Shepher and Lionel Tiger review what they consider the relevant literature on female hierarchies and examine a particular case: the formation of hierarchies among women working in the kitchen of an Israeli kibbutz. They come to a similar conclusion concerning females in our own species. The female hierarchies they found were "problematic structures" characterized by "problems of discipline, reluctance to accept authority, and rather strained relations among the workers." Furthermore, there was no indication of any female hierarchy for the kibbutz as a whole. In sum, the authors could find scant evidence for any formalized competition among women.[42] My own survey of the literature leads me to the same conclusion: with very few exceptions, studies in the social sciences say almost nothing about competition among women; certainly, no firm data have been collected.[43]

We are confronted with a peculiar situation. It is hard to imagine that anyone who considers the evidence summarized in this chapter can continue to accept the stereotype of a female primate so absorbed in rearing infants that she is unable to involve herself in the social organization of the group. Once we start to look closely at primates, there is not a single precedent for such a female. Obviously, the grain of truth in the stereotype remains the universal commitment of female primates to reproduction. But they clearly have an equally powerful and universal commitment to compete, and in particular to quest for high status. Access to resources—the key to successful gestation and lactation—and the ability to protect one's family from members of one's own species are so nearly correlated with status that female status has become very nearly an end in itself.

From the nearly solitary galago, who excludes every female but an occasional daughter from her range, to a troop of several hundred hierarchically arranged rhesus macaques; from the tiny marmoset, who suppresses breeding in her female associates, to a tightly knit sisterhood of langurs guarding their feeding grounds from neighboring sisterhoods and protecting their nurseries from marauding males: the central organizing principle of primate social life is competition between females and especially female lineages. Whereas males compete for transitory status and transient access to females, it is females who tend to play for more enduring stakes. For many species, female rank is long-lived and can be translated into longstanding benefits for descendants of both sexes. Females should be, if anything, *more* competitive than males, not less, although the manner in which females compete may be less direct, less boisterous, and hence more difficult to measure.

Here then is the puzzle. Competition between females is documented for every well-studied species of primate save one: our own. Once we leave the scientific realm, of course, and consider history, literature, and, for many of us, personal experience, examples of highly competitive, manipulative, and even murderous females flock to mind. Whether they are competing for inheritances, resource-controlling "eligible" husbands, real power in the form of a crown or a partnership, or simply "points," one scarcely need search far to find famous examples. The Empress Livia, Lady Macbeth, Mrs. John Dashwood, and Strindberg's two genteel ladies who ever so subtly put one another down while conversing sedately in a Stockholm café—these happen to be my favorite examples. But returning now to science: where is the hard evidence for these competing ladies? Without it, we remain in the realm of fiction and opinion, and this world does not lack the critics who will promptly tell us so.

The solution to this problem of the "missing species" is hazarded, then, as an opinion only, my own: Women are no less competitive than other primates, and the evidence will be forthcoming when we begin to devise methodologies sufficiently ingenious to measure it. Efforts to date have sought to find "lines of authority" and hierarchies comparable to those males form in corporations. No scientist has yet trained a systematic eye on women competing with one another in the spheres that really matter to them.

The difficulty is not simply narrowness of vision and the mistaken assumption that female competition will take the same form as competition between males, but also the subtlety of interactions between females. Consider the problem that a human ethologist would face in an effort to measure quantitatively such phenomena as sisters-in-law vying for a family inheritance which is to be passed on to their respective children, or the competition for status between mothers who perceive, however dimly, that their own "place in society" reflects on the whole family and that it may determine the rank at which their own children enter the community at maturity. The quantitative study of such behavior in a natural setting hardly exists. We are not yet equipped to measure the elaborations upon old themes that our fabulously inventive, and devious, species creates daily. Ethologists who have to contend with malarial swamps and subjects that hide in dense foliage 40 feet above them confront a task simple by comparison. How do you attach a number to calumny? How do you measure a sweetly worded put-down? Until we are able to solve such problems, evidence for this hypothesized competitive component in the nature of women remains anecdotal, intuitively sensed but not confirmed by science.

Yet the problems are more than scientific ones, because whether or not we can document the natural competitiveness of women, we are still going to have to deal with the fairly well documented problem that unrelated women have working together over a long period of time—even when they share a common goal. Feminists like Carolyn Heilbrun have repeatedly lamented "the failure of women to bond." Historians who have traced the course of women's movements, and the history of women's role in politics, echo the same refrain. "Women everywhere," writes the historian William O'Neill, "seem to identify more with their families than with their sex and join men in supporting role definitions that keep women out of politics. This lack of solidarity distinguishes women from the minorities with which they are often compared. Racial and ethnic groups vote in blocs and support their candidates, giving them a leverage that women politicians do not have." Masculine prejudice is only one element of those obstacles which have kept women from direct political involvement. "Behind the question of why women seldom run for office lies the more far-reaching question of why they do not identify with, and are not loyal to, their own sex."[44] It is my own belief that an answer to this question will come only when we begin to consider women from a perspective some millions of years longer than most historians usually have in mind.

Women are really dreadfully complicated aren't they . . . I mean,
with other animals, well the majority of them, they're either off heat
or on heat. Everyone knows where they are. I probably should have
been born a horse or something.   ALAN AYCKBOURN, 1975

7

The Primate Origins
of Female Sexuality

Moulay Ismail the Bloodthirsty, former emperor of Morocco.
Shinbone the Yanomamo chief. Plain old Clifford Curtis of
Hudson, Maine. Each has achieved a twofold immortality—
once in print and again in their dozens of offspring (respec-
tively 888, 43, and 32 direct descendants).[1] This litany of potent names provides documentation for the widespread conviction that a male's reproductive capacity is enormous,
infinite almost, compared to the limited reproductive potential
of a female. In contrast to Shinbone, a wildly successful
woman might leave two dozen offspring—at the outside. In
most traditional societies, five or fewer would be more like it.

Much has been made of this basic difference between the
sexes. In polygynous breeding systems, reproductive variation
between a successful Don Juan and a male who never reproduces ranges from zero into the hundreds; the variance of reproductive success between females is much smaller. But this
truism has led to a widespread misconception concerning the
nature of females.

Because of the focus upon males and upon the impressive
variation in reproductive success between one male and another, the comparatively small variation between females has
been treated as insignificant or even nonexistent.[2] The emphasis has been on "competitive males" and "nurturant females." Consider the cream of the crop of recent textbooks on

sociobiology: "Most adult females in most animal populations are likely to be breeding at or close to the theoretical limit of their capacity to produce and rear young. Among males, by contrast, there is always the possibility of doing better."[3] Since a male's reproductive success is thought to be dependent upon his readiness or eagerness to copulate with any fertile female (if the risks are low enough), natural selection would have favored sexually assertive males. But females, all of whom (it is assumed) were already breeding close to their reproductive potential, would find no advantage in large numbers of copulations, since most of the copulations would not result in conception.[4] Consequently, sexual assertiveness in females would not evolve—or so runs the argument.

The sociobiological literature stresses the travails of males—their quest for different females, the burdens of intrasexual competition, the entire biological infrastructure for the double standard. No doubt this perspective has led to insights concerning male sexuality. But it has also effectively blocked progress toward understanding female sexuality—defined here as the readiness of a female to engage in sexual activity.

There are at least two serious misconceptions in this extremely male-focused view of sexuality. The first is the assumption that all females in a natural state—unlike males—breed at or near their reproductive capacity, and that consequently there is little room for natural selection to operate on females. In fact, there is considerable variance in the reproductive success of females. Although data over a lifetime for female primates in a completely natural state are nonexistent, short-term data (under ten years) for both wild and provisioned nonhuman primates, as well as data from "primitive" or preindustrial human societies, make clear that females differ in fertility. Perhaps more importantly, the mortality of immatures—which is and probably always has been high—varies from mother to mother.

Nancy Howell, who followed the reproductive careers of 166 !Kung women over 11 years, reported that the maximum number of births for any woman was nine, the minimum zero; the modal number of children born was five. Using a computer-simulated model, she calculates that "52%

of a cohort of women will fail to have any children . . . An additional 5% will fail to have any grandchildren."[5] Tentatively (because her sample sizes are necessarily small) Howell concludes that long birth spacing among the !Kung (three to four years) may be attributed to the time it takes a woman in this foraging society to recoup fat reserves expended in growth, lactation, and pregnancy. In a comparison between bush-living women—that is, women living more or less as their ancestors did—and !Kung women married to husbands working near Bantu cattleposts, Howell found that the generally fatter cattlepost women whose diets contained substantial portions of milk and grain tended to have shorter birth intervals.[6]

Among the Bedik people of eastern Senegal and among the Mende of Upper Rural Bambara Chiefdom, Sierra Leone, the number of children born to monogamously married women tends to be greater than that of women married in a polygynous union.[7] Barry Isaac's study of 160 rural Mende women showed the average fertility of a monogamously married woman was 4.3 children, compared with 3.7 for a woman married polygynously. What was most fascinating to me, however, was the unusually high fertility of the senior wife or "big wife" in polygynous unions, which was even higher than that of monogamously married women. Isaac suggests that the high status of senior wives within the polygynous household may allow them a better diet, a lower level of physical exertion, and a lower level of psychological stress than that enjoyed by either their junior co-wives or lone wives in monogamous unions.[8] If true, this is one of the rare examples in the ethnographic literature which fits patterns emerging for females in other species of primates: differential use of resources and differential fertility according to rank.

The invention of food storage and exchange systems, as well as the many other forms of insurance that humans devise to guard against famine, probably means that interfemale variation in fitness was greater among prehominids than among contemporary humans, just as it appears to be greater for living nonhuman primates. The link between rank, nutritional status, and reproductive success is fairly well demonstrated for species such as wild toque macaques in Sri

Lanka.[9] Fifteen percent of Wolfgang Dittus' toque macaque
study population was wiped out during a period of food shortage; this and other studies indicate that mortality falls most
heavily upon animals already under physiological stress—
subordinates, lactating females, and especially immatures,
85 percent of whom died before maturity in the Toque case.[10]
Such mortality means heavy selection pressure not just upon
males but upon the mothers of hapless young of either sex.
During periods of political instability and frequent male takeovers in species where infanticide occurs, infant mortality in a
troop over a period of years can run as high as 80 percent or
more.[11] Mothers may differ substantially from one another in
the survival of their offspring, and any behavior that improves
the chance of survival of her infants will be favored by natural
selection. As we shall see, sexual assertiveness may have
functioned in just this way.

This brings us to the second problem with the view that
sexual assertiveness was not adaptive for females: the assumption that copulations serve no function other than insemination. This assumption, coupled with the fact that a female's
theoretical capacity to conceive is quite limited compared with
a male's theoretical capacity to inseminate, and that a female
would not lack for males to mate with, have led to the conclusion that natural selection would not favor the evolution of a
sexually assertive female. But how, then, does one explain the
sexual appetite of the human female, who has the capacity to
engage in sex on virtually any day of the month, at any time
of the year? Her sexual receptivity is clearly not synonymous
with ovulation; if it will not lead to conception, what is all this
nonreproductive sexuality about?

The most common explanation for "continuous receptivity"
and for other womanly attributes such as prominent breasts
and buttocks has been that they evolved among humans to
make a woman permanently attractive to her mate. One of the
earliest and most widely read spokesmen for this view was
Desmond Morris in his 1967 book, *The Naked Ape*, an account of "the sexiest primate alive." Morris argued that the
development of continuous receptivity, breasts, buttocks, and
orgasm in women was crucial to the evolution of human

beings because they cemented the pair-bond between mates by providing mutual rewards for both sexual partners.[12] The reduction of human body hair and the unusually large size of the human penis compared with that of chimpanzees and gorillas are explained by similar arguments about mutual attraction.[13]

This sexual bond was regarded as essential for enlisting the male's help in rearing highly dependent human babies. Over a decade later, it is still widely believed that the female of our species became uniquely sexualized in order to attract and keep her man, and that continuous receptivity, along with orgasm, developed in the human female to ensure that she would be willing to copulate frequently enough to satisfy her man's needs, and that it would be to her camp he would return from the hunt.[14] An extension of this view is that willingness to copulate evolved in women as a "service to males,"[15] or—an even odder corollary—that the female orgasm evolved "to make it easier for the female to be satisfied by one male."[16]

There can be little doubt that the widespread acceptance of the pair-bonding hypothesis derives from the subjective experience of generations of western scholars. Few could comfortably belittle the mutual attachment so readily observable between mated primates—particularly between human sexual partners, where the emotions of lovemaking and companionship are magnified by countless preadaptations for intelligence, subtlety, sharing, and loyalty, as well as for heightened mutual pleasure from sexual intercourse. And there is substantial objective evidence from several quarters that accords rather well with the notion that increased female sexuality promotes monogamous bonding and male investment in offspring.

Masters and Johnson, for example, have recorded increased libido in women during the first and particularly second trimester of pregnancy,[17] a finding which fits the prediction generated by the Morris hypothesis that a female in imminent need of male assistance will exhibit increased receptivity. Furthermore, subcultures with an emphasis on marital stability and constancy tend to have more elaborate sexual foreplay.[18]

One can also point to parallels between continuous receptivity in humans and the pattern of sexual activity among such monogamous primates as marmosets, who are characterized by substantial investment by males in the offspring of their mates. As in humans, the early phases of marmoset pairing are marked by frequent copulatory bouts. Established partners then settle into a pattern of copulation at a low level throughout the menstrual cycle, with occasional peaks in sexual activity around the time of ovulation (which in rapidly breeding marmosets often occurs shortly after the mother gives birth). If anything, ovulation among marmosets is harder to detect than it is in humans, since marmosets do not menstruate.[19]

Such evidence strengthens the likelihood that some connection exists between pair-bonding, paternal investment, and the evolution of continuous female receptivity—or what might more precisely be termed "situation-dependent receptivity," since clearly none of these creatures is continuously engaged in sexual behavior. But the theory that female humans became *uniquely* sexualized in order to strengthen the pair-bond and thereby increase paternal investment in offspring has been weakened not only by the marmoset evidence but by other evidence about the sexual behavior of primates.

Before elaborating on these recent studies, I wish to stress at the outset that there is not a single species of primate for which adequate data on reproductive strategies exist. Thousands of hours of observation of captive, wild, and urban primates still leave us short of definitive answers. At best, then, this chapter can aspire to replace conventional wisdom with plausible hypotheses about female sexuality which, though still speculative, will be better aligned with available evidence and less narrow in perspective.

The picture is incomplete, but already it is apparent that primates generally are far more sexualized than we had imagined. The readiness of female primates to engage in sexual activity at times when conception is out of the question, or to engage in more sexual activity with more partners than is necessary for conception, needs an explanation. While a compelling case can be made for the old hypothesis that the readiness, even eagerness, of women to engage in sexual activity both at ovulation and at other times as well evolved in order to elicit male investment in offspring, it is clear that this theory is insufficient by itself to explain the undeniable readiness of females in a wide array of other primates to engage in nonreproductive copulations.

Findings for other hominoids do not challenge the view that lovemaking is especially intricate in the human case— any more than the precocity of highly trained chimpanzees who are able to communicate with symbols challenges the linguistic preeminence of the human species. But such findings caution us against viewing as unique qualities capacities that are far better understood as matters of degree. Caresses, face-to-face coupling, and prolonged mutual gazing in a sexual context can be seen in rudimentary form among both wild and captive great apes. Copulations witnessed by Biruté Galdikas among wild orangutans in central Borneo tended to be face-to-face, with one or both partners hanging from an overhead branch. Thrusting lasted 3 to 17 minutes and was almost always preceded by the male's oral contact with the female's genitalia. Mature males utter a long grumbling call just before, or during, intromission.[20] Mutual stimulation of the partner's genitalia has been observed most often among captive great apes (gorillas, orangutans, common chimpanzees, and pygmy chimpanzees), where observation conditions are better and perhaps, too, where the animals are bored. Among pygmy chimpanzees, for example, either partner may use one foot to manipulate the scrotum or vulva of the other. Most remarkable is the copulatory gaze described for this species: partners stare intently into each other's eyes.[21] Even the smile and the kiss, staple gestures of flirtation and foreplay among humans, have precursors among primate expressions of submission and conciliation, the wide-mouthed grimace, touching of hands, embracing, the grazing of lips.[22]

There is far more here than the "slam-bam, thank you ma'am" which was the prevailing summation of primate sexual relations in the early seventies when I was a graduate

Chimpanzees

student. There can be no doubt that the sexually sophisticated human species remains at the highly erotic end of the continuum, and our position there does indeed demand an explanation—but not necessarily a special explanation applicable *only* to humans. We are an extreme case, not a separate one. In seeking to understand human eroticism, one does well to keep in mind an image that, like all good science fiction, is at least a possibility: future generations of chimpanzees transmitting among themselves (in sign language) a rudimentary version of the *Kama Sutra*.

A<small>MONG</small> primate females, approximately midway through each menstrual cycle an egg cell is released from one of the two ovaries and travels down the fallopian tubes. In humans, and most other species, the menstrual cycle is about 30 days, but for some species cycles can be as short as 7 days or as long as 60. In a small proportion of human females, release of the ovum is accompanied by a perceptible pang known as *mittelschmerz*. For most women, however, the time of ovulation is marked by a scarcely perceptible rise in temperature and a rise in estrogen and other hormones that only laboratory analysis can detect.

Women may engage in sexual activity on any given day of the reproductive cycle even though taboos and injunctions against intercourse at particular times (such as during menstruation or postpartum) are found in many cultures. Though statistics on the frequency of intercourse are rare for any species, a recent sample for one of this planet's better studied populations reveals that married women in America have intercourse on average about ten times a month, though the variation is enormous.[23] Several studies report a greater likelihood of sexual interactions around midcycle, though the issue is extremely controversial.[24] In 1978 a team of psychologists— David Adams, Anne Burt, and Alice Ross Gold—studying married women in a college community in Connecticut reported a statistically significant peak in sexual behavior around the time of ovulation. Differences between this study and earlier ones which had expressly stated that no peak occurred at midcycle were attributed to research designs which permitted male-initiated sexual encounters to obscure the natural cyclicity of feminine libido. Instead, the new survey focused on sexual behavior initiated by females or else instigated by *both* partners. The report indicating that human females were especially receptive at midcycle elicited a flurry of dissenting commentary.[25]

Despite the controversial nature of the Adams findings, they are in accord with a recent study among !Kung San (Bushmen) women in the Kalahari desert. The !Kung San study combined in-depth interviews of San women with analysis of hormone levels in blood samples which were taken at the same time. The results of this collaborative effort by Carol Worthman, Marjorie Shostak, and Mel Konner showed a statistically significant peak in sexual behavior at midcycle, with an accompanying peak in rate of orgasm. The interviews revealed an increase in sexual relations with lovers as well as husbands during this period.[26]

If indeed an increase in libido at midcycle does occur among women, our species parallels other primates in this respect. Most monkeys and apes exhibit a moderate to sharp rise in sexual activity around the time of ovulation. In nonhuman primates living in multimale troops, this peak in sexual receptivity is typically much sharper and is often conspicuously signaled by edematous, burgundy-colored swellings

about the vulva. Gross morphological changes are accompanied by a transformation in behavior known as estrus. During
the rest of the month, and during the long periods when she
is either pregnant or lactating (and hence not cycling), this
female would rarely engage in sexual behavior. But when in
estrus, a baboon, for example, may copulate as often as 100
times during a single estrous period. In a ten-hour segment of
this period she may copulate as often as 23 times with three
different males.[27] These bouts of frantic sexual activity are energetically costly and quite perilous in terms of both predation
and injury by conspecifics. The Greek term *estrus* literally
means gadfly and carries with it the implication that females
are driven to distraction by this temporary buzzing in their
endocrine system.

In part owing to historical accident—because baboons were
extensively studied at an early date—species such as baboons,
mangabeys, and chimpanzees which happen to exhibit gaudy
sexual swellings at midcycle became widely regarded as standard models for primates as a whole. Typically, such primates
would be contrasted with humans, where there are no visible
signs of ovulation. Because it was assumed that sexual swellings and strictly cyclical sexual behavior were characteristic of
all pre-human primates, the conventional questions were:
Why was estrus lost in the course of human evolution? Why
did ovulation become concealed? Why did human females become continuously receptive?

It was in this context that Desmond Morris and others constructed the hunting hypothesis and their model based on the
need for a pair-bond to ensure the survival of especially helpless human young.

> First, he had to hunt if he was to survive. Second, he had to have
> a better brain to make up for his poor hunting body. Third, he
> had to have a longer childhood to grow the bigger brain and to
> educate it. Fourth, the females had to stay put and mind the
> babies while the males went hunting. Fifth, the males had to co-
> operate with one another on the hunt. Sixth, they had to stand
> up straight and use weapons for the hunt to succeed . . . Females
> had to develop a pairing tendency [and] if the weaker males were
> going to be expected to cooperate . . . they had to be given more

sexual rights . . . Each male, too, would need a strong pairing tendency.[28]

For Morris and others, this is where loss of estrus and the development of continuous receptivity came into the picture—as uniquely human adaptations to strengthen the pair-bond and, in the process, strengthen the group. As Jane Lancaster puts it, "What would happen to the division of labor if human females came into estrus? If times are bad and vegetable food scarce, who is going to go hunting if there is an irresistible female in camp? But more important, what happens to the special relationship between mates under these circumstances? The human adaptation has been the suppression of estrus and continuous sexual interest under conscious control of both partners."[29]

But newly available information about the patterning of sexual activity among monkeys and apes raises serious doubts about the pair-bond explanation for female sexuality. First, research on monogamous primates (with the exception of marmosets) reveals a poor correlation between the existence of pair-bonds and high levels of sexual activity between mates or of sexual activity across the cycle. Second, though some primates fit the estrous model, others are not so easily categorized as cyclical; recent observations of monkeys and apes make it seem highly unlikely that supposedly "unique" human attributes such as concealed ovulation, nonreproductive sexual activity, and female orgasms (as we will see in Chapter 8) are in fact unique to women.

The assumption that continuous receptivity evolved to cement the pair-bond is opened to doubt by observations of the sex lives of such monogamous primates as siamangs, gibbons, and indri. In the year that Jonathan Pollock spent in Madagascar observing *Indri indri,* he did not observe a single completed copulation. Similarly, gibbons confine breeding to periods a few months in duration which may occur only once in two or three years. During these breeding periods, sexual activity is limited to one or two copulations a day in the case of white-handed gibbons, or to one copulatory bout every other day in the case of siamangs.[30] Yet, couples observed over a pe-

riod of seven years were apparently mated for life. Male gibbons, then, are tolerating hiatuses in sexual activity of several
years' duration without abandoning their mates.

On the basis of such information, it has become increasingly common among primatologists to scoff at the conventional explanation for human sexuality. These skeptics argue
that if paternal care is so critical for infant survival, fathers
would be selected for paternalism, and that sexual appetite
among males would be selected against. A siamang father
would scarcely need to be seduced by his mate into carrying
their juvenile offspring: his own reproductive success, as well
as hers, would be in jeopardy if he neglected to do so.

Furthermore, several behavioral biologists have argued that,
far from increasing reproductive success, constant sexual activity would interfere with such vital pursuits as defense of
territory or accumulation by females of the bodily reserves
needed for gestation and lactation. To counter such objections, Richard Alexander and Katharine Noonan have suggested that what strengthened the pair-bond between early
human men and women was not frequency of copulations,
but rather the fact that ovulation became concealed. Without
estrus, the only guarantee a male would have of consorting
with a female near the auspicious moment of ovulation would
be to consort with her throughout the month, month after
month. Concealment of ovulation "enabled females to force
desirable males into consort relationships long enough to reduce their likelihood of success in seeking other matings, and
simultaneously raised the male's confidence of paternity by
failing to inform other, potentially competing males of the timing of ovulation." The ability of females to conceal ovulation
would be even more effective if women themselves were unaware of when they ovulated; no intimate confidence or subconscious gesture would betray the physiological event.

Alexander and Noonan stipulate a specific social environment for this transition from advertised to concealed ovulation: mixed-sex communities in which females were not totally inaccessible to males other than their mates. They argue
that "concealment of ovulation could only evolve in a group-
living situation in which the importance of parental care in

offspring reproductive success was increasing, and that these two circumstances together describe a large part of the uniqueness of the social environment of humans during their divergence from other primates."[31]

An even more intriguing hypothesis to explain the evolution of concealed ovulation—one certain to be controversial—has been offered by the zoologist Nancy Burley. Burley explains concealed ovulation as a corollary of a specifically human attribute: possession of an intellectual capacity sufficient to manipulate fertility, artificially, through contraception. She argues that human females would have consciously limited the number of children they produced if they could have, and this would have stopped short of the maximum number of which they are capable. While avoiding pain at childbirth, and the risk of complications, as well as the extra workload of another infant, such women would necessarily have left fewer descendants. Hence, natural selection would have countered these conscious desires for fewer babies by endowing women with reproductive cycles recalcitrant to birth control. Modern medical specialists find it difficult to detect the precise time of ovulation. Could a preoccupied gatherer, then, consistently second-guess her ovaries to avoid copulations on the days when she might conceive?[32]

Although Burley's arguments will seem counterintuitive to those familiar with the satisfactions of bearing and raising children, she marshals a variety of supportive evidence from the ethnographic literature. In particular, she cites several cultures in which women explicitly state that if they could choose, they would have fewer children, and she points to a nearly universal skewing in the incidence of pronatalism: husbands, parents, or in-laws are frequently more eager to see additional children born than the women who has to bear and care for them.

Either hypothesis—Burley's suggestion that cycles refractory to birth control have been selected for, or Alexander's argument that continuous receptivity and nonreproductive sexuality force a male to focus on a single mate—could explain the remarkable difficulty of detecting ovulation in our species. But neither hypothesis can explain concealment of ovulation and assertive solicitations of males by nonovulating females among primates which are *not* pair-bonded, not living in a communal context, not sufficiently self-aware for birth control to be a possibility.

And this brings us to the second problem with the pair-bond theory: nonreproductive, situation-dependent (rather than strictly cyclical) sexual receptivity is not uniquely human; it has been reported under both ordinary and extraordinary conditions among a variety of primates. Those exhibiting occasional noncyclical receptivity include species which are polygynous and lacking substantial male investment (such as langurs) as well as species that live in multimale troops (vervets). Furthermore, even among species where either seasonal or cyclical receptivity is the rule (as among savanna baboons or barbary macaques), females exhibit an aggressive sexuality—by human standards, nymphomaniac—which goes far beyond the necessary minimum for ensuring insemination. In contrast to what one might expect if female sexuality arose in order to enhance pair-bonds, such assertive sexual behavior with numerous males may be most pronounced among species that live in multimale troops. Whereas some distinctly pair-bonded primates (like the gibbon) copulate only during discrete intervals which occur every two or three years, a virtually solitary ape like the orangutan may copulate at any time in the reproductive cycle; for these species, female sexuality is clearly not linked to monogamous pair-bonding.

Primate sexuality is far more complex than anyone was suggesting ten years ago, and certainly more complex than any unitary explanation based on human pair-bonds can comfortably encompass. In particular, numerous exceptions to the anthropological adage that nonhuman primates confine their matings to midcycle can be documented among the higher primates. Unquestionably, other primates tend to be more seasonal in their breeding activities than humans are, and peaks in intensity of sexual solicitations are more jagged, but nevertheless, as has been pointed out by several authors—in particular Frank Beach and Thelma Rowell—there has apparently been in the course of primate evolution a trend, or more probably several trends, away from strictly seasonal and cyclical determination of receptivity.[33]

Most prosimians or "lower primates" exemplify the classic mammalian mode of strictly seasonal receptivity with discrete breeding days. Among the nocturnal African galagos, for example, mean cycle length is 44 days with an estrous period (solicitations) lasting 12 days. Within these 12 days females actually permit intromission on perhaps 6 of them. As among other mammals, copulations do not occur outside of these circumscribed periods, and periods of receptivity themselves are seasonal, occurring once or perhaps twice during the year. Most of the year, no sexual activity takes place. By contrast, higher primates tend to be more flexible in their breeding patterns, although this is not to say that exceptions cannot be found. For example, rhesus and Japanese macaques breed only during specific months, one season a year. This season is virtually the only time that females ovulate, and male sperm production tracks this schedule of female receptivity.[34] Nevertheless, during the breeding season, copulations may be witnessed on any day of the cycle, and in this respect macaques comply with a general trend of increasing flexibility.

For the purposes of general discussion, the patterning of sexual activity among primates can be classified as follows: Some genera such as the baboons (*Papio*) or the mangabeys (*Cercocebus*) are strictly cyclical, although the precise length of the cycle can be extended or compressed by social circumstances. These genera tend to contain terrestrial troop-dwelling species with multimale breeding systems; female receptivity is advertised by conspicuous sexual swellings.[35] At the other end of the continuum, primates such as orangutans and marmosets are fairly continuously receptive throughout the cycle and, like humans, do not advertise receptivity with any conspicuous morphological displays.

The majority of higher primates falls somewhere in between these two extremes. That is, they are generally cyclical but have the potential to lapse into situation-dependent sexual receptivity *not* predetermined by day of the menstrual cycle. In many of these species there are no visible signals of estrus other than behavioral presentations to males—such as the female langur who presents and shudders her head. Most colobines, for example, lack sexual swellings, as do some cercopithecines. Macaques are characterized by variable swellings and changes in the color of sexual skin, but except for a few species, these changes are not necessarily correlated with ovulation, and some of the most striking color changes occur during pregnancy. In many species, females continue to exhibit estrous-like behavior or "pseudo-estrus" for several cycles after conception.

For most species in this intermediate category between strictly cyclical receptivity and receptivity which is situation-dependent, exact endocrinological and social circumstances surrounding the transition to situation-dependent receptivity are unknown. At least one precipitating factor has been identified, however: the appearance of unfamiliar males. A captive chimpanzee who has been failing to cycle may develop large sexual swellings overnight if she is moved near to a male. Wild langur females will actively solicit invading males who have just entered the troop, presenting to these new males, shuddering their heads, and copulating with the males day after day—even if the females are already pregnant. Among patas monkeys in an experimental colony, the incidence of pseudo-estrus among pregnant females similarly increased when a new male was introduced.[36] Such instances represent lapses of cyclicity under circumstances which are probably fairly unusual in the life of the animal. But semicontinuous or continuous receptivity may sometimes be normative, as has been shown in several studies involving macaques and gelada baboons.[37]

The weak correlation between ovarian state and sexual behavior has been most often noticed among captive animals belonging to three species of great apes: orangutans, common chimpanzees, and particularly pygmy chimpanzees.[38] Gorillas in captivity appear to be more cyclical; they may also be cyclical in the wild, though it is rare to see gorillas copulating under natural conditions.[39] Information gleaned from studies of captive animals is problematic, however, particularly when males and females are caged separately and re-introduced to one another for daily observations. Sexual behavior in these instances may be complicated by the re-establishment of relationships between the two animals. For this reason, information from the wild, however limited, is especially valuable.

On the basis of matings between 16 pairs of wild orangutans witnessed by Biruté Galdikas in the course of 41 months of observation in the jungles of Borneo, it appears that young adult male orangs who sometimes accompany otherwise solitary orang females will attempt to copulate—often by force— with their traveling companions regardless of the stage of the female's reproductive cycle. Consortships with the more dominant, fully adult males, however, often involve female initiative, the female actively seeking out and sexually soliciting these males.[40] Although evidence from wild orangutans accords fairly well with observations in captivity, the same is not true for common chimpanzees, who apparently are more cyclical in the wild. Caroline Tutin reports on the basis of 1,002 copulations observed for seven wild chimpanzee females at the Gombe Stream Reserve that copulations occur at high frequencies during the period of maximum swelling and at "low-to-negligible" frequencies at other swelling states.[41]

Even in the most cyclical primates such as baboons, gorillas, or chimpanzees, or in the most seasonal of breeders such as macaques that mate only during a few months of each year, the amount of sexual activity far exceeds the minimum necessary for insemination. A female chimpanzee or baboon may exhibit a strong preference for one male rather than another and establish a special consort relationship with that male, but she will nevertheless mate in the course of the days surrounding ovulation with several or even many different males. Thirty-five percent of estrous females in a troop of savanna baboons studied by Glenn Hausfater at Amboseli had different male consorts in the morning and afternoon of the same day during those four days presumed by Hausfater (on the basis of visible swellings) to be near the time of ovulation.[42] Caroline Tutin reports that wild female chimps in estrus alternate between consortships with a single male in the privacy of the forest, out of sight from other animals— what Gombe researchers refer to as "on safari"—and group matings involving coteries of males. Males in these clusters sometimes fight, but there is little advantage to them from doing so since other would-be suitors move in to profit from the distraction. Whereas a female on safari may copulate 5 to 10 times in a day, an estrous female traveling with a group of males may copulate 30 to 50 times in a day.[43] Gorilla females are typically found in one-male breeding units but may nevertheless solicit and mate with subordinate "blackback" males in addition to the dominant "silverback" male in groups where such young males happen to be present.[44]

The most seemingly "promiscuous" of all primate females have to be the Barbary macaques. These *Macaca sylvanus* represent an isolated North African branch of an otherwise strictly Asian genus; they are confined to the inhospitable climes of Morocco's Atlas mountains, except for a small outpost of the species still maintained today by the British on the Rock of Gibraltar. According to the primatologist David Taub, a Barbary macaque at midcycle copulates on average once every 17 minutes, and mates at least once with every sexually mature male in the troop. Taub characterizes sexual relationships in this species as "fluid, brief and temporary." Each day during a midcycle period of estrus, a female solicited, established, and terminated numerous sexual relationships. In the majority of cases, the female was the sole initiator of the relationship and exercised considerable control over the situation. Taub's data show that on any given day, an estrous female would mate with a third to nearly all of the 11 adult and subadult males on each of the several days that she was in estrus. A female would spend a brief period with one male during which one copulation typically occurred, then deliberately abandon him and immediately solicit another male.[45]

Clearly, estrous behavior around the time of ovulation may be directed toward many more partners than are necessary for conception. Even odder, females may be especially assertive when they are unlikely to conceive or when they could not possibly conceive—for example, pseudo-estrous behavior during pregnancy. Among langurs, pregnant females may repeatedly solicit and copulate with alien males who invade the troop. Even without precipitating social circumstances, female primates continue to exhibit some degree of sexual receptivity through the first months of pregnancy.[46] Among both wild chimpanzees and savanna baboons, the duration of estrus is longest, and in some cases the sexual swelling is most pronounced, among young females, many of them still in the period of adolescent sterility. Similarly, wild rhesus macaque females in their first or second season of breeding were the most obvious and assertive in their sexual solicitations.[47]

Despite indisputable differences between the intense sexual activity of an estrous monkey and the more subdued and subtle solicitations of women, human females cannot properly be said to hold a monopoly on either nonreproductive sexual activity or on noncyclical situation-dependent sexuality. It is possible, of course, that all this estrous brouhaha is due to endocrinological accident, that it is merely a by-product of hormonal fluctuations in a complex system, that it evolved for reasons unrelated to the reproductive strategies of these permanently social animals, unrelated to the lifetime fitness of females. But is it probable that seemingly unproductive estrous bouts and solicitations in excess of those necessary for fertilization would crop up by accident repeatedly and take different forms in different lineages? Could bouts of unproductive estrus be retained in different genera—and in spite of obvious costs—unless somehow the fitness of females was affected for the better? Obviously a leading question. It marks the transition in this chapter from a historical and factual review to my own opinions.

CONSIDER the alternative possibility, that nonreproductive sexuality in females is adaptive, that there is more to female sexuality than simply ensuring that sperm is internally introduced at about the time an egg is released. This of course is not a new idea. It was this argument—that there is more to sexuality than insemination—which originally led to the hypothesis that continuous receptivity evolved to enhance pair-bonds and ensure paternal investment. But a misplaced emphasis on the difference between human and nonhuman primates, and a tendency to view evolution from a male perspective, meant that the case was shut too soon. Other implications of continuous receptivity were left unexplored.

Granted it is advantageous for a pair-bonded female or a female in a harem to copulate with her mate, but is it also advantageous for her to solicit outside males? When is it advantageous for her to mate with more than one male? When does it behoove a polygynous female to mate at times when she could not possibly conceive? There are many ways in which females might benefit from mating promiscuously with more than one partner.

One possibility is that the female increases her ability to choose a male with superior genes to father her offspring. A female who establishes consortships with many males will be better able to assess the capacities of each and, perhaps more importantly, might be in a position to engineer the situation so that she consorts with a particular male (whether he is currently the dominant male in the group or not) on the day of ovulation. Elaborating on the model of sexual selection, Robert Trivers has proposed that females enhance the fitness of their offspring by thus selecting males with superior genotypes.[48] Hence, a female paired with a genetically "inferior" male could benefit from copulating with her mate during pseudo-estrus and soliciting an outside male nearer the time she actually ovulates. For species where ovulation is truly concealed—even from the females themselves—this would obviously not be an option. Furthermore, there are no data for any primate species that either support or refute such arguments. At best, there is anecdotal evidence for a few species, such as savanna baboons, that females occasionally prefer to spend time with one male prior to conception and another male after giving birth during the lactation period.[49] This, however, is a far cry from the fine-grained discriminations Trivers seems to have had in mind.

Whether or not females can assess genotypes, it might be to a female's advantage to have her offspring fathered by a dominant male. It is known that on the first days of her sexual swellings, a female in a multimale troop copulates with a number of different males, including low-ranking males and even subadults. As her swelling reaches maximum turgescence, or deepest color, about the time of ovulation, she may form consortships with one or more dominant males who are able to defend their right of access to this maximally turgescent female in a way that a subordinate male could not. Two zoologists studying elephant seals, Catherine Cox and Burney LeBoeuf, have suggested that by signaling her receptivity to a broad array of males, a female would incite competition among bulls.[50] Under such conditions, only a male able to defeat other males in the ensuing competition would be able to mate with her. Other zoologists, in particular Tim Clutton-Brock and Paul Harvey, have applied similar arguments to the sexual swellings of primates.

Clutton-Brock and Harvey point to the fact that conspicuous swellings advertising ovulation tend to be found among species arranged in multimale breeding systems.[51] Sexual swellings are commonest among baboons, mangabeys, and macaques and may also be found among talapoin monkeys and several New World monkeys such as howlers.[52] Only two species of colobines, olive and red colobus monkeys, exhibit swellings, and at least one of these (red colobus) is known to live in large, multimale troops, an unusual social arrangement for colobines. Audible calls given by females during copulations are another trait which might advertise ovulation and thereby incite competition between males. Interestingly, the staccato grunts emitted by one or both partners during copulations are more complex and presumably noticeable among multimale baboons than they are among monogamous gibbons or married human couples.[53] One interpretation of this finding is that such signals are adapted for communication with animals *outside* of a particular consortship, not with the partner.

One difficulty with extrapolating to monkeys from a model originally devised with elephant seals in mind is the difference between the social system of these large marine mammals and the breeding arrangements of higher primates. Whereas herds of elephant seals come together on land only once each year, most primates live together year-round. Furthermore, although baboon and macaque males do migrate between troops, they tend to do so gradually. The vicissitudes of male dominance are less drastic among multimale species than among harem dwellers with frequent male takeovers. Primate females in multimale troops would scarcely need an up-to-the-minute report concerning the relative merits or ranking of males who are already well-known to them.

There can be little doubt that swellings visible at considerable distance increase competition for ovulating females. But did such swellings evolve to provide females with information about male quality or dominance or to ensure that she mate with a dominant animal? Consider another hypothesis: that it is not choosing one ideal partner, but rather attracting a number of partners in a limited time period, that is the point of advertising receptivity in multimale troops. By attracting the attention of a number of males, it is energetically less costly for a female to engage in multiple matings. Instead of the female's having to solicit each male in turn, it is the male who approaches her and who must either create the opportunity to copulate with her (by threatening or fighting another male) or forsake foraging and other activities to follow her over a period of time, seeking an opportunity for clandestine matings out of sight of more dominant animals. According to this hypothesis, then, sexual swellings evolved in multimale troops not to provide the female with information about which male was dominant or to ensure that she mate with that particular male, but rather to ensure that she mate with a *range* of males. Because dominant males tend to monopolize females around the day of ovulation, they may have the highest probability of inseminating her, but not the only probability.

A fairly good case, I think, can be marshaled in favor of the hypothesis that sexual swellings evolved to ensure the formation of multiple consortships. An alternative, but certainly not mutually exclusive, possibility is that these swellings serve also as diplomatic passports permitting females freedom of movement beyond the confines of their normal range under the protection of interested males.[54] It is worth noting, therefore, that three of the four species in which adult females as well as males sometimes migrate between communities are also species characterized by sexual swellings. Female transfer has been reported among South American howler monkeys, African red colobus, and chimpanzees; all three exhibit conspicuous swellings at midcycle.[55] Females also move between troops among gorillas, but this species does not signal ovulation with swellings.

Conspicuous sexual swellings and a willingness to copulate with many males guarantee females in many species a certain freedom of movement and range of choice. But what about the initial question: Why multiple males?

Different primates vary enormously in the amount of care that males provide infants, but in all species of the order the behavior of males has an important effect—be it positive or negative—upon the survival of infants which goes far beyond mere contribution of sperm. Throughout evolutionary history, there would have been intense selection pressure against a male who attacked or, in the case of those species where male investment is crucial to infant survival, who ignored his own offspring. The uncertainty which inevitably surrounds paternity favors any female able to plant a seed of doubt. Even if a male is not a sufficiently probable progenitor to induce him to invest directly in an infant by caring for it, if he has mated with that female, it is unlikely that he could rule out completely the possibility that he fathered subsequent offspring. By pushing a father toward the conservative edge of the margin of error that surrounds paternity, a female may forestall direct male interference in her offspring's survival.

Females in different species vary greatly in the options open to them for manipulating the information about paternity that will be available to males. Females with conspicuous swellings around ovulation save themselves some time and trouble; on the other hand, only males in the group at the time of a female's swelling (unless she moves between groups at this time) are likely to be caught up in her advertising campaign. For females among such species as langurs, marmosets, and humans, who live under the fairly constant surveillance not of several males but of one male—a harem leader or a husband—opportunities for copulations with multiple males would be less frequent. But when they occur, these females might find it to their benefit to take advantage of them. Under these circumstances, conspicuous swellings advertising ovulation would be a disadvantage, since they would invite increased surveillance by the female's mate at the time of ovulation and might limit to this short period her attractiveness to other males. For these females, the concealment of ovulation provides considerably greater flexibility in the timing of their copulations with outside males.[56] And again, the advantage to a female of soliciting outside copulations might be to encourage males to behave "paternally" toward her infant should they later have contact with it. Especially where infanticide is a significant hazard, as among languis in one-male troops and chimpanzees in multimale troops, cycling or already pregnant females invest in the survival of future offspring by initiating consort relations with a variety of males, including both the resident male or males and potential invaders or usurpers.

This hypothesis assumes, of course, that males remember the identity of past consorts. Is this overtaxing monkey intelligence? Evidence from rodents who are surely, if anything, less intelligent than primates are, suggests not. Two Canadian zoologists, Frank Mallory and Ronald Brooks, devised carefully controlled experiments to study the impact of prior exposure to the mother on the behavior of male lemmings toward her infants. Male collared lemmings from the Canadian arctic were added to a cage containing a mother with her new litter. For 16 of the litters, the newcomer was the "stud male" who had fathered the litter and then been removed from the cage right after mating; none of the infants in these litters was harmed. However, in the 32 litters exposed to unfamiliar males—males who had never mated with the mother—42 percent of the pups were killed. Typically, the female attacked the male as soon as he was released into her cage; if she reduced the intensity of her attack, the male killed her offspring by biting them in the head. By contrast, "when the stud male was introduced to the [familiar] female's cage, he was also attacked immediately; however, once contact had been made, fighting usually stopped and after a short sniffing bout, the female returned to the nest. The stud male often helped care for the young."[57] Among rodents, copulation cannot occur after insemination; hence, mock estrus is not an option. Nevertheless, the zoologist Jay Labov has devised an ingenious experiment to learn if mating with an impostor can forestall infanticide. Labov coated a sexually receptive (ovulating) house mouse with urine from a pregnant donor. Males introduced to the donor and her subsequent litter behaved like "stud males" if they had mated with her impostor. Males who had not mated with the mother's olfactory impostor were more likely to kill the mother's pups.[58]

Returning to primates, we might ask why males, in their turn, would copulate with nonovulating females. Why have not males been selected to discriminate between ovulation and pseudo-estrus? There are several possible answers to this puzzle. In the first place, in species such as langurs, some males *are* more discriminating than others. A troop leader may ignore the solicitations of a female who is actively sought after by extratroop males. One could argue that the troop leader is jaded by multiple opportunities to copulate, but it seems more likely to me that his nonchalance is due to the fact that he is in a position to continuously monitor the reproductive solicitations of the same female. Individual variation between females in intensity of solicitations and pheromones emitted might make it difficult to accurately gauge the timing of ovulation except through long-term comparison of the female's signals from one day to the next. Hence, males may possess some capacity to discriminate between ovulating and nonovulating females but they would not always have sufficient information available to them to make the distinction.

Another possibility would be that there is some fractional chance that a male can induce ovulation in a female who does not happen to be at midcycle by copulating with her. Induced ovulation is known to occur in some mammals (such as rats) and is suspected to sometimes occur in primates (such as humans). That is, although ovulation would usually occur spontaneously around midcycle, in some cases it could be brought about by copulation; this fractional chance of insemination might make it worthwhile for subordinate or extratroop males to copulate when there was the opportunity.[59]

One of the objections to the hypothesis that females benefit their offspring by engaging in promiscuous matings has been the assumption that mates of longstanding may *withhold* investment from offspring born to promiscuous females. Hence the situation would appear to be more complicated in the case of monogamous primates where there might be the danger of a cuckolded male deserting. In the case of species such as marmosets where male care is essential to survival of the twins, this would be reproductively disastrous for the female. But it is important to keep in mind that any male in a monogamous species could afford to desert only if he actually stood a good chance of locating another suitable territory and finding another mate. He could of course attempt to chase out his unfaithful mate, but only at considerable risk since she will typically be equal in size and by disposition not easily daunted. Better perhaps to sequester his mate as best he can, and thereby reduce the likelihood of her mating with any other male. Under most circumstances, his chances of being the progenitor still remain better than even.

T HERE may be a great deal of method in the gadfly-madness that gets classified as "estrus." Sexually assertive females draw into the web of possible progenitors several—even many—different males. In some cases, males may be inhibited from harming infants by this fractional chance of paternity; in others, males may be induced by a likelihood of paternity to make positive contributions toward the survival or the social status of females and their infants.

There is wide variation both between and within species, particularly the human species, in the degree to which males care for infants. Not surprisingly, males most likely to invest appear to be those in situations providing them with a relatively high certainty of paternity, such as monogamously mated marmosets or siamangs, or, in the case of polygynous species (several species of macaques, for example, or savanna baboons), those males who have engaged in a sustained consort relationship with a particular female. Under such circumstances, even males in multimale social systems sometimes exhibit considerable devotion in the form of carrying immatures and protecting them from predators, particularly conspecifics.

Consider the Barbary macaque, a species with multimale troops characterized by extraordinarily high rates of copulations and seeming promiscuity. According to David Taub, males regularly interact with infants in a variety of caretaking activities during the infants' first year of life. Most males direct their attention toward one or two specific infants, so that each infant is cared for by several different males—only one of whom can actually be the father. Taub argues that such male attention is essential for infant survival. In support of his argument, he notes that the single case of infant mortality during his study involved "the one and only infant that never interacted with adult or subadult males."[60]

Taub has attempted to reconstruct the evolutionary pressures that bias females toward a communal caretaking strategy rather than a single consort, paternal caretaking system: "Although it is unknown whether care from only one male would be insufficient for infant survival or for obtaining other benefits, the amount of care received from just the one male would clearly be considerably less than the total care received by the other infants, each of whom would receive attention from several different males. If it is assumed that the benefits of male-care to an infant are proportional to the number of males contributing to it, then it follows that the reproductive success of the deviant female would be less than that of a female using the common strategy [where each offspring would receive care from several males] ."[61]

By actively seeking out multiple males, females offer all of them some probability of having sired her offspring; as Taub points out, this may be the most effective means for females to obtain the cumulative investment from several males that her offspring will need to survive.

I N ORDER to breed successfully and rear young, female primates must overcome substantial challenges. These include competition with other animals for resources, the threat that interference from adults of either sex—particularly males—poses to the survival of offspring, and even competition between females for males, or at least for the services males have to offer. Competition for resources (or for the status that entitles her to resources) can have considerable effect on a female's production of offspring and their survival, but her ability to manipulate the behavior of other animals, especially males, is no less important. A female's success at tapping the services of males or forestalling males from harming her offspring will often depend on her capacity to forge sexual relationships with these males, and thereby confuse the issue of paternity. For males, sexual behavior may often be little more than a matter of insemination, but for females, sexual consortships and mating have far wider implications— most especially in the primate order where male behavior has such a crucial bearing on the survival of infants.

The readiness of females to engage in nonreproductive, or what might even be called "extrareproductive," sexual activity is clearly not an exclusively human attribute. A tendency away from strictly cyclical receptivity can be found among higher primates generally, and it is characteristic of both species with and those without pair-bonding, and of species both with and without extensive male investment. This is not to say that new and special dimensions of pair-bonding did not attend the hominid transition to a gathering-hunting way of life. These arguments do not challenge the old idea that the prolonged helplessness of human young placed strong selective pressures upon males to help rear their young and on females to seek to ensure such paternal assistance. But it is increasingly apparent that situation-dependent receptivity (and with it concealed ovulation) as well as nonreproductive sexual activity generally, predated the revolutionary new lifestyle we associate with early humans—a life centered around a home base, characterized by division of labor into gathering and hunting and by an emphasis on the exchanges of food and services between mates.[62] A highly assertive female sexuality marked by a potential to shift from cyclical to situation-dependent receptivity constituted the physiological heritage that prehominid females brought to this evolutionary experiment. What happened to this increasingly cerebral female and to the sexual legacy she carried with her from her primate past is the subject of the next chapter.

I [Boswell] argued that the chastity of women was of much more consequence than that of men, as the property and rights of families depend upon it.

Surely [said Boswell's ladyfriend] that is easily answered if a woman does intrigue but when she is with child. [Boswell could not answer this, but it made him feel "uneasy."]

DIARY OF JAMES BOSWELL, 1774–1776

8

A Disputed Legacy

An assertive, temporarily insatiable female sexuality, epitomized by the seemingly nymphomaniac solicitations of a Barbary macaque in estrus—what earthly relevance does the conduct of this monkey have for understanding her culture-bearing cousin, whose solicitations are sedate, self-conscious, often elaborate in their subtlety and indirection? We can imagine sound evolutionary reasons for Nature to cultivate an assertive sexuality in her prehuman daughters—we examined some of these reasons in Chapter 7—but to trace natural selection's legacy to the present, through the intervening biological and historical vicissitudes, is even more difficult than making an after-the-fact inventory of the contents of an intricate estate which at different times has been owned by a variety of people. Some of its heirs might wish to know which are its initial assets and which were more recent acquisitions. Which assets were abandoned along the way as ill-suited to prevailing conditions? The heirs here, of course, are women, and the legacy is the biological infrastructure of their sexuality.

May it please the court, there are circumstances in this case which are unusual. Ownership of the legacy is not in question. Disputation arises when we attempt to determine just exactly *what* is owned, and how it was acquired. A more awkward litigation can scarcely be imagined. First, the court will be asked to admit evidence for events which have no surviving witnesses or documentation. Surrogate depositions will be taken from witnesses who watched distant relatives. We are asked to regard living nonhuman primates as stand-ins for our common ancestors who lived twenty million years ago, and the perils of this assumption must be acknowledged. It is entirely feasible to examine clinical evidence for female sexual responsiveness in other primates, but when we extrapolate from these laboratory depositions to the lives of primates who lived in the forests and savannas where the relevant transactions actually took place, the record goes hazy. Halfway through the litigation, the court recorder is asked to skip across vast tracts of evolutionary time and to substitute for hard evidence about our surrogate ancestors a haphazard collection of ethnographic and historical hearsay about our actual ancestors and very near relations—accounts which describe the way woman's sexual proclivities have been perceived in various societies. But it has to be recognized that such accounts do not necessarily tell us anything about the legacy itself, woman's sexuality.

Then we close the depositions and turn to a title search through the fossil evidence. These proceedings are even more disappointing than makeshift testimony and hearsay: the record is mostly blank. The record for the longest and most critical period in the whole litigation, the period from twelve to five million years before the present, is absent entirely. Even when our genealogy becomes populated with some of the better known hominids, beginning about five million years ago, the most we can say about these owners of the legacy, the australopithecines, is that they walked on two feet but otherwise resembled chimpanzees: their brains were chimp-sized, they had large front teeth, and the disparity in size between the two sexes was at least as great—quite probably greater—than is the case in contemporary human populations.

Jurors, you may grumble at the poverty of such evidence and at the awkward transitions from monkey facts to human ideologies. But recognize that the alternative is to abandon hope altogether of establishing what the legacy consisted of. Above all, keep in mind, lest you leap to false conclusions, that no party in this litigation is suggesting that people are nothing but sophisticated monkeys. Culture-bearing monkeys maybe, but that seven letter word spells a world of difference.

"Sex," Bronislaw Malinowsky wrote after he had traveled among Pacific Islanders, "is not a mere physiological transaction . . . it implies love and love-making; it becomes the nucleus of such venerable institutions as marriage and the family; it pervades art and it produces its spells and magic. It dominates in fact almost every aspect of culture."[1] Yet it stretches the fabric of culture beyond its true limits to assert, as another very eminent anthropologist, Clifford Geertz, has, that sexuality is merely "a cultural activity sustaining a biological process."[2]

For women, the physiology of sex is, to be sure, profoundly influenced by socially conditioned expectations, their attitudes toward men and marriage, and their self-image. Even responses such as sexual climax, which ought to be physiologically straightforward, are contingent in real life upon individual attitudes and cultural practices. Arapesh women, for example, rarely experience orgasm—no concept of a female's sexual climax even exists for them—whereas among the Mundugumor or Samoans the woman's orgasm is considered a routine part of sex.[3] A single culture—our own, for example—can change rapidly. Consider the pendulum swing in less than three decades between the women reported on by Alfred Kinsey in 1948 and those surveyed by *Cosmopolitan* magazine in the fall of 1980. In Kinsey's study, 3 percent of girls had had intercourse by age 15. By age 24, 8 percent of young wives had engaged in extramarital affairs; the incidence rose to 20 percent by age 35. Slightly less than half of the women surveyed reported that they usually had orgasms, 8 percent reported that they never did. Of the women who responded to the *Cosmopolitan* questionnaire in 1980, 20 percent had had intercourse by age 15. Among married women 18–34, 50 percent had been unfaithful, while among women older than 35 the rate of infidelity rose to nearly 70 percent. For those having intercourse on a regular basis, 60 percent responded that they usually or always achieved orgasms; only one percent responded that they never did. The most striking change in the last thirty years has been in the number of lovers a woman is likely to have. Some 30 percent of the *Cosmopolitan* sample have had 2 to 5 lovers; 10 percent have had more than 25.[4]

Few generalizations about sexuality apply cross-culturally. But one of the few that does hold is that couples engaged in sexual activity tend to seek privacy. To get their information, ethnographers and social scientists must rely mainly on interviews, often with self-selected volunteers or paid informants. Only very recently has anyone collected rigorous data amenable to verification by repeating the study in a laboratory. No one has undertaken a study in which women were followed for years and asked about their sexual experiences. Even the suggestion to participate in such a study would probably be shocking to a substantial portion of the very cross-section of humanity whose participation would be essential for validity of the study. In sum, we have no information about the sexual lives of women comparable in its validity to what is known about savanna baboons.

When a paucity of information is combined with poignant interest, controversy is the inevitable result. So it is scarcely surprising that the few authors writing in any detail about the evolution of women's sexuality have come to radically different verdicts. At one extreme, it has been argued that all women have inherited a strong, innate sexual drive from their primate forebears; the only reason this fact of life is not acknowledged is that male-dominated, patriarchal cultures suppress it to make their control of women more effective. At the other pole, it has been argued that sexual feeling in women has no evolutionary importance. Rather, male sexual drive has been critical in human evolution, and desire in females is merely vestigial, a by-product of the masculine phenomenon.

Mary Jane Sherfey, a feminist psychiatrist, addressed women's sexuality in attempting to bring a revised Freudian theory in line with more recent findings of sex research and biology. The cornerstone of her evolutionary argument was the assertion that "throughout primate evolution, selective pressure has always tended in the direction of favoring the development of the longer duration of the intense orgasmic contractions in the females" in order to "remove the largest amount of venous congestion in the most effective manner."[5] This emphasis on the medical value of the orgasm has a certain venerable provenance. It can be traced at least to Galen in the second century, and it has persisted, passing in and out of favor, since then. Catholic women in seventeenth-century France were permitted by medical opinion to masturbate to orgasm should the husband ejaculate and withdraw before she was satisfied—but not as a concession to pleasure. Rather, it was believed that her orgasm would contribute to producing strong, healthy children.[6]

Apart from this vague notion that orgasms are therapeutic, Sherfey offered no evolutionary rationale for her proposition that "to all intents and purposes, the human female is sexually insatiable in the presence of the highest degree of sexual satiation."[7] Her ideas were enthusiastically endorsed by some feminists and social scientists, ignored or rejected by others.[8] Her reconstruction of human evolution, which was definitely idiosyncratic, fueled the criticisms. She accepted the radical feminist vision, adapted from nineteenth-century antecedents, of a matriarchal stage in human evolution. This brazen age was supposedly an era in which women engaged unrestrainedly in sexual activity. To maintain this hypothesis, however, it is necessary to brush aside the troubling fact that there is virtually no archaeological or ethnographic evidence for it.

Lacking support from either biology or the social sciences, Sherfey's evolutionary views have not caught on, and her belief in the fundamental sexual nature of women, though not necessarily wrong, has been neglected. One of her harshest critics (but also someone who has read her work carefully) is Donald Symons, an anthropologist-turned-sociobiologist, who has recently written:

> It is difficult to see how expending time and energy pursuing the will-o'-the-wisp of sexual satiation, endlessly and fruitlessly attempting to make a bottomless cup run over, could conceivably contribute to a female's reproductive success. On the contrary,

insatiability would markedly interfere with the adaptively significant activities of food gathering and preparing and child care.
Moreover, to the extent that insatiability promoted random matings, it would further reduce female reproductive success by sub-
verting female choice.

Symons' case rests on the conclusion that "in a natural habitat, females appear to vary relatively little in the number of
children they produce during their lifetime" and that females
cannot improve their statistics "by copulating with many
males."[9] The strict measure of a woman's evolutionary success will be relatively constant in a given environment and
not much affected by her sexual conduct. Therefore, her sexuality is immaterial to her evolutionary history.

Why then do women ever experience sexual desire, much
less orgasm? Symons argues that women have sexual feelings
for much the same reason that men have nipples: nature
makes the two sexes as variations on the same basic model.
From this perspective, female orgasms occur as "a by-product
of mammalian bisexual potential: orgasm may be possible for
female mammals because it is adaptive for males."[10]

If Sherfey's ideas smack of Galen and of France under
Louis XIV, Symons' has antecedents not only in Aristotle but
in the early nineteenth-century denial that women have any
sexual drive whatever. The Victorian belief, which gained cre-
dence just as evolutionary thought was also taking hold, was a
reversal of the view prevailing just a century earlier: that
women had an "exuberant and inexhaustible appetite for all
variety of sexual pleasure." By the second half of the nineteenth century, the popular medical authority William Acton
could assert with confidence that "the majority of women
(happily for them) are not much troubled with sexual feelings of any kind," and as late as 1906 the doctrine persisted
that "in many [women] the appetite never asserts itself."[11]

The notion that woman's orgasm is "in an evolutionary
sense a 'pseudo-male' response"[12] appears to be a vestige of
Victorian thought on the subject. But it is buttressed by a
kind of evolutionary utilitarianism that has a more modern fla-
vor: female orgasm is unpredictable, unreliable. An organ—or
in this case a physiological mechanism—that is important to

fitness (or reproductive success) and properly adapted to its task should work better than that. As evidence in support of this argument, it is often argued that the orgasm in human females is an evolutionary oddity—a relatively new phenomenon arising from the heightened sexuality of human males and not a legacy from their remote maternal ancestors. Animal females, it is said, do without.

Physiologically detectable, psychologically impressive, the female orgasm has become the central issue in a long debate about the nature and evolution of female sexuality. We need to consider this psychophysiological response in some detail.

Nobody really denies that orgasm occurs sporadically among women. Margaret Mead pointed out the differences among both individuals and cultures in this regard. She referred to the climax as a "potentiality" not always realized.[13] Yet it is clear that the capacity for orgasm is universal. One of the insights gained from the research of Masters and Johnson is that virtually all women, sufficiently prepared and stimulated, do have orgasms, but not necessarily from intercourse or from intercourse alone.

Surveys conducted by Seymour Fisher, Shere Hite, and others indicate that for the majority of American women clitoral stimulation—manual, oral, or otherwise—is necessary for orgasm, and that intercourse does not suffice in the absence of clitoral stimulation. From these surveys, it appears that only about a quarter of women regularly climax from intercourse alone.[14] Consequently, Hite characterizes "conventional" orgasms, achieved through indirect stimulation of the clitoris during intercourse, as, at best, a "Rube Goldberg scheme"—on the whole impractical.

In strictly anatomical terms, then, the human clitoris is the core of the problem. Evolution is virtually always a compromise between preexisting structures and selection favoring improvement. Yet even as a compromise, women's sexual machinery seems peculiarly inefficient. On the other hand, it is not a new invention; it is widespread among mammals and nearly universal among primates (though with widespread variation in size, design, and placement). It is inconspicuous in the tree shrew but clearly detectable in lemurs, lorises, and virtually all the higher primates. It may be somewhat more prominent in species such as baboons and macaques which tend to live in troops with many males, but otherwise the clitoris is surprisingly variable in species of Old World monkeys. Among some baboons, the fold of skin surrounding the clitoris expands into a pendulous lobe during estrus. In the apes, particularly the gibbon and chimpanzee, the clitoris is well developed; it is larger than the human organ, both absolutely and relative to body size.[15]

No function other than sexual stimulation of the female has ever been assigned to the clitoris,[16] and this is very likely the reason that it has been a subject either tabooed or ignored in textbooks, even very recent accounts.[17] Are we to assume, then, that this organ is irrelevant—a pudendal equivalent of the intestinal appendix? It would be safer to suspect that, like most organs—including even the underrated appendix—it serves a purpose, or once did. But the purpose, as noted above, appears to be transmitting the pleasurable sexual stimulations that sometimes culminate in orgasm. And that brings us full circle: to rationalize the existence of a clitoris in evolutionary terms, we must show that female orgasms confer some reproductive advantage on the creatures experiencing them.

Orgasm was originally a term for any intense excitement, or merely a dispassionate noun meaning inflammation or swelling, as in either a wound or a ripe fruit. Nowadays, when applied to American women, orgasm refers to a highly variable, typically pleasurable psychophysiological phenomenon. At its culmination, a local build-up of blood in the vessels is released, and muscular tension developed in response to sexual—most often clitoral—stimulation is relaxed. Most of what we know about female orgasms comes from rather recent research by Masters and Johnson and others, from interviews with articulate groups of Western women, and from sparse investigations in other societies.[18] Cross-cultural study of the subject is bedeviled by problems of translation; experienced anthropologists are often unsure what term connotes orgasm, or whether such a word even exists. Whether nonhuman females experience something corresponding to orgasm is also difficult to ascertain—and there are no words with which to put the question directly—but at least animals can be directly observed in a way that human females rarely have been.

Both wild and captive female primates of many species experience "orgasms," if we accept the opinions of many of the researchers who have studied them in the wild. But what is the evidence? Certain responses to copulation—such as spasmodic arm movements, staccato grunts, and lipsmacks which could merely indicate general excitement—are routinely reported. "Trancelike expressions" sometimes follow self-manipulation (a phenomenon that may depend on the eye of the beholder). And specific physiological responses have been recorded: rhythmic contractions of the vagina and changes in heart rate, which correspond to similar observations in human females. Unfortunately, the data are precise only when the animals are removed from their natural surroundings and studied in white-walled laboratories. The experimental conditions themselves may be stressful and abnormally stimulating.

All in all, the straightforward question, "Do other primates experience orgasms?" yields a mixed bag of answers. A poll, probably not exhaustive, of published opinions on the subject turns up two groups of authors. One of them has written that female orgasms are uniquely or primarily a human experience. In includes Desmond Morris, David Barash, George Pugh, Frank Beach, Richard Alexander, Katharine Noonan, and Donald Symons. Another group (which, interestingly, contains more women) opines that nonhuman females do have such orgasms. Among its members are Frances Burton, Elaine Morgan, Suzanne Chevalier-Skolnikoff, Jane Lancaster, Doris Zumpe, Richard Michael, Donald Goldfoot, and myself.[19] This state of affairs should caution us to pay close attention to the evidence and its limitations, and to distinguish between what is known, what is experienced, and what is believed.

Among rhesus macaques, a female at the time of her partner's ejaculation turns her head and looks back toward the mounted male; she reaches toward him with spasmodic arm movements known as a "clutch reflex." In 1968 the psychiatrists Doris Zumpe and Richard Michael tentatively suggested that this reaction might be "an external expression of consummatory sexual behavior in the female rhesus monkey."[20] Zumpe and Michael monitored the copulations of three nonpregnant rhesus females. Of 389 copulations that culminated in ejaculation, 97 percent were associated with a clutch response. Frame-by-frame analysis of films taken during the copulations shows that the clutch reflex begins while the male is still thrusting. This sequence suggests that the female's clutch response itself may trigger ejaculation. The clutch response depends on normal levels of estrogen, the female hormone; it is suppressed by ovariectomy and restored when estrogen is administered.

More physiological evidence for orgasm in nonhuman primates was collected several years later in experiments undertaken by the anthropologist Frances Burton. In her experimental trials, Burton subjected three rhesus females to five minutes of grooming, five minutes of clitoral stimulation mechanically applied by the experimenter, four minutes of rest, and five more minutes of vaginal stimulation.[21] Each of the monkeys clearly exhibited three of Masters and Johnson's four copulatory phases: the excitement phase characterized by vaginal dilation, mucous secretion, and labial engorgement; a plateau phase with clitoral tumescence; and the resolution phase involving detumescence of the clitoris. As with the human subjects, specific signs of the third "orgasmic phase" could not be detected because the clitoral glans is obscured during this phase.[22] Burton concluded from her experiments that rhesus do indeed possess orgasmic capacity, but she acknowledged a major difficulty in interpreting her results: the actual duration of each copulatory bout among rhesus monkeys under natural conditions is very short, about three or four seconds. Levels of stimulation comparable to those which induced orgasm in the laboratory would occur in the wild only if there were multiple copulatory bouts and sexual stimulation was cumulative from bout to bout.

At least the first set of circumstances—repeated copulations with partners in a very short time—are observed among macaques, baboons, and chimpanzees living under natural conditions. Yet, no one can say what levels of sexual stimulation are achieved in the wild, or how the animals experience them. In the human female, orgasm may be greatly facilitated by high arousal prior to intromission; arousal may accumulate over a period of hours,[23] and women, unlike men, do not return to a physiologically unaroused state after orgasm but to the preorgasmic level. In short, although there is no way to measure the level of sexual arousal these wild primates experienced during hours and days of sexual activity, there exists a distinct possibility that stimulation sufficient for orgasm occurs in the wild.

The occurrence of orgasm in wild primates even seems probable if we allow ourselves to accept less rigorous specifications than those set forth by Masters and Johnson. Behaviors readily apparent to a fieldworker, who cannot assess specific physiological responses, are suggestive indeed. "Copulation calls," that is, the series of staccato grunts given by a wild baboon or macaque female at about the moment her partner ejaculates,[24] involuntary muscle spasms affecting the whole body or portions of it,[25] or a pause after sexual activity which is accompanied by particular facial expressions and panting[26] have all been construed as signs of the female's climax. Such signals may accompany heterosexual or homosexual interactions. Homosexual activity involving two females has been reported for a variety of wild and captive primates.[27]

While observing a captive group of stumptail macaques, Suzanne Chevalier-Skolnikoff saw one female mount another on 23 occasions. Although female monkeys may mount others as an expression of dominance (just as males mount males), sexual stimulation can be a prominent feature in such interactions. Chevalier-Skolnikoff detailed the course of events of such mountings. Typically, the soliciting female approaches another with a teeth-chattering expression, then the approached female presents her rump: "Soliciting female mounts presenting female and, over a period of about one minute, executes a series of pelvic thrusting movements, thereby rubbing her genitals on the back of her mountee. [The mounter is stimulated, but the mountee is not.] Both fe-

Stumptail macaques: round-mouth look and female mounting male

males make puckered-lips, lip-smacking, or square-mouthed expression." Then, in what Chevalier-Skolnikoff refers to as the "orgasmic phase," the "mounter pauses, and for about ten seconds, manifests muscular spasms, accompanied by the frowning round-mouth look and the rhythmic expiration vocalization."[28]

Stumptail macaques observed in a laboratory by D. A. Goldfoot and his colleagues confirm the impression that females engaged in either homosexual or heterosexual activities experience some sort of sexual climax. These experimenters were able to establish that the female's round-mouthed face (which in a male signals ejaculation) coincided with intense uterine contractions and a sudden increase in heart rate. Four of ten females observed during heterosexual copulations displayed the round-mouthed response during at least one copulatory episode; on average the female would make the face 10 times in the course of 52 tests, but females were very individual in their responsiveness. One of them made the "ejaculation face" in 40 percent of her sexual encounters.[29]

Free-ranging Japanese macaques living on a ranch near Laredo, Texas, resemble their stumptail cousins in their sexual behavior. (Their troop has been named Arashiyama West after the original primate colony of that name in Japan.) Because the monkeys are unrestrained, mechanical devices were not inserted to measure their internal physiological events. Even so, we can say that the females unambiguously seek the same sort of genital stimulation that led to orgasm in the laboratory. A third of all observed matings were preceded by the *female* mounting her partner several times to rub her perineal region against his back. As the observer, Linda Wolfe, points out: "Thigh pressure and rubbing would, of course, facilitate orgasm if these females possess the capacity for orgasm. That adult females who mounted males were more sexually motivated is borne out by the finding that those adult females who mounted males had a statistically significant greater number of male partners . . . than those females . . . who did not."[30]

A conservative interpretation of all these findings would be that under some circumstances female primates find genital stimulation positively reinforcing. Orangutan and chimpanzee females living in the forest as well as captive monkeys and apes will, on occasion, provide it for themselves, although masturbation among females is observed far less commonly than it is among males.[31] Still and all, it seems unlikely that primate females "are not much troubled with sexual feelings of any kind." What, then, are we to make of this libidinous aspect of their nature?

ALTHOUGH evidence is increasing that orgasms do occur in other primates, no really plausible explanation for their purpose has been forthcoming. None of the various physiological or therapeutic rationalizations (relieving congestion of blood, promoting fertilization)[32] has received much support from clinical research. The lack of obvious purpose has left the way open for both orgasm, and female sexuality in general, to be dismissed as "nonadaptive," "incidental," "dysgenic," or adaptive only insofar as it provides a service to males.

Sherfey began her investigation from an equally plausible postulate: that genital pleasure is as adaptive in females as it is in males, that it creates an inclination to seek partners and persist in copulation until satisfaction is obtained. But adaptive for what?

The point of copulation, most biologists would hasten to point out, is insemination. Because a female need be inseminated only once to establish a pregnancy, it would seem that there could be little advantage to her from repeated copulations, and none whatever from copulation when she is not ovulating. This objection leads with apparent inevitability to the conclusion that female sexuality is an evolutionary vestige.

Yet only a failure to think seriously about females and to consider the evidence would allow someone to conclude that natural selection operates more powerfully on male sexuality than on female sexuality, or to believe that the female's reproductive character could be "invisible" to natural selection. When some variability occurs in the members of a species, natural selection has an opportunity to choose between them, no matter how slight the variation. And for females, faced with the doubly difficult adaptive task of gaining access to resources in the environment, converting them into offspring, and seeing that those offspring survive, the stakes are very high, every bit as high as for males. There is clear evidence that females of many species take advantage of numerous opportunities to augment their chances for reproductive success. For example, one female may forestall reproduction in another—what might be called the "Hagar phenomenon": a socially dominant female suppresses ovulation in her subordinates or excludes a rival from safe harbors and feeding sites, just as Abraham's wife, Sarah, drove her husband's concubine, Hagar, into the wilderness. For another example, females take advantage of high social rank, which entitles them and their offspring to nutritious food, caretaking, and protection—factors that matter a great deal in the survival of offspring. A female who lacks protection and support may fail to raise a single offspring to breeding age, even though she gives birth repeatedly. We now know that females are not breeding machines that automatically produce one offspring after another from menarche to menopause; in fact, only a portion, perhaps a small one, of females living in nature could even approach the theoretical maximum number of offspring (see Chapter 7). A mother's social circumstances have far-reaching consequences: the availability of food and helpers, protection from predation and from other members of her own species, all of these are critical for both her and her offspring. In such a world, selection could hardly be operating only on males, and it is exceedingly unlikely that the energy consumed by female sexual activity, or the risk it entails, could have persisted unless it somehow enhanced a female's reproductive success.

If we recognize that a female's reproductive success can depend in critical ways on the tolerance of nearby males, on male willingness to assist an infant, or at least to leave it alone, the selective importance of an active, promiscuous sexuality becomes readily apparent. Female primates influence males by consorting with them, thereby manipulating the information available to males about possible paternity. To the extent that her subsequent offspring benefit, the female has benefited from her seeming nymphomania.

Under some circumstances, of course, sharing the pleasure of intercourse with a particular male might reinforce her ties to him, but enjoyment of sex could also serve her interests at the expense of the male's. "Insatiability"—Sherfey's term—is probably too strong a word, but an inclination in the appropriate situation to solicit males could serve an important adaptive purpose in the lives of female primates. Sherfey's premise that the main features of woman's sexual anatomy evolved in a context where selection favored prolonged solicitations involving multiple partners no longer looks so far-fetched.

Mary Jane Sherfey, it should be remembered, was writing in the early sixties without benefit of the post-1965 explosion in knowledge about primate behavior. Although she is not explicit on this point, readers should be aware that she bases her model of primate sexuality on the behavior of only a few species—rhesus macaques, savanna baboons, and chimpanzees, all of which live in polygynous breeding systems. She did not consider, for example, nonhuman primates living in monogamous families. But Sherfey's main focus was on the human case, and as luck would have it, her errors of omission do not necessarily jeopardize her arguments. None of the great apes—indisputably our closest living relatives—are obligate monogamers, and evidence of pronounced sexual dimorphism in fossil anthropoids and hominids makes it unlikely that any of our immediate ancestors were either.

Rarely can paleontology provide insights into social structure, but on this particular point—whether or not a species possessed a monogamous breeding system—I think we are on fairly firm ground. Recall that most primates in which females are equal in status to males are monogamous, and virtually all monogamous species are characterized by monomorphic, or same-size, sexes. By contrast, polygynous species tend to be characterized by sexual dimorphism, with males substantially larger than females, owing to the fact that males in polygynous species fight among themselves for access to harems of females. In monogamous species there exist approximately equivalent levels of overt intrasexual competition in both sexes.

When we look at the fossil record in the hominid line, the degree of sexual dimorphism among our antecedents was at least as great as that in contemporary human populations, where men are 5 to 12 percent larger than females; recent fossil evidence from Hadar in Ethiopia and Laetoli in Tanzania provides grounds for believing that hominids four million years ago were even *more* dimorphic than humans are today. Certainly this was the case for Old World higher primates in the more distant past, at thirty million years before the present. When we compare our statistics to those for mammals generally, *Homo sapiens* falls into the range of a "mildly polygynous species." Among our highly polygynous hominoid relatives, the gorillas, chimpanzees, and orangutans, males tend to be 25 percent (or more) larger than females.

Despite these findings about sexual dimorphism, the old idea that early humans were monogamous continues to attract proponents. However, taking this position now necessitates a certain anthropocentrism and special pleading. For example, in his otherwise very innovative and comprehensive essay "The Origin of Man," Owen Lovejoy argues that pair-bonding and a monogamous breeding system were crucial to the emergence of modern man; but to do so he must first downplay the evidence for dimorphism in early humans by arguing that "human sexual dimorphism is clearly not typical." He stresses the fact that human canine teeth are not dimorphic, as is the case in other sexually dimorphic, polygynous animals. But this is a weak link in his argument, since it does not seem necessary to assume that early man used his teeth in fighting as other primates do. If a club better served his purposes, there would no longer be selection for male canines.[33]

The paradoxes of human sexuality—the mismatch between men, who are transiently impotent after an orgasm, and women, who are not only capable of multiple orgasms but may prefer them[34]—may not be so paradoxical after all, if we no longer assume that these traits evolved in a strictly monogamous context. The physiology of the clitoris, which does not typically generate orgasm after a single copulation, ceases to be mysterious if we put aside the idea that women's sexuality evolved in order to "serve" her mate, and examine instead the possibility that it evolved in order to increase the reproductive success of primate mothers through enhanced survival of their offspring.

But even so, the conundrums hardly disappear. If we assume that women have been biologically endowed with a lusty primate sexuality, how have cultural developments managed to alter or override this legacy? Has women's sexual behavior been permitted to drift from its biological moorings? Is it no longer subject to natural selection? Must we assume that behavior which was once adaptive is no longer adaptive? How has women's sexuality changed in the intervening five million or so years since we shared an ancestor in common with the chimpanzees?

How do we even approach such questions? Rarely does a human society permit women the sexual independence that, say, macaques have. If there were no such thing as a "compromising" situation, what would women do? Reports of women engaging in intercourse with a series of partners are few indeed, and it is hard to know how to interpret them.[35] What we know of primates suggests that prehominid females embarking upon the human enterprise were possessed of an aggressive readiness to engage in both reproductive and nonreproductive liaisons with multiple, but selected, males. What happened next is, and probably will always remain, shrouded in mystery. We can only document the attitudes prevalent in human cultures during historical times. There can be no doubt from such evidence that the *expectation* of female "promiscuity" has had a profound effect on human cultural institutions.

T wo conflicting idealizations—woman as chaste, passive, sexually innocent, and woman as possessor of a dangerous sexuality—always have dominated historical assessments of female nature. Similar tensions characterize the beliefs in many preliterate, traditional societies.[36] Female sexuality is much on people's minds. Whether in village gossip or television soap operas, the affairs of women are a matter of compelling interest.

Almost universally, sexual sanctions are stricter for women than for men. As the anthropologist Alice Schlegel has pointed out, nearly twice as many human cultures forbid adultery by wives as those that proscribe it for husbands.[37] Even societies which appear to esteem women for their sexual purity and passivity nevertheless take extensive precautions to prevent them from breaching their chastity. On one point there is an extraordinary consensus: woman's readiness to engage in sexual activity is great enough that the majority of the world's cultures—most of which determine descent through the male line—have made some effort to control it. The reason for expending all this effort usually comes down to some variant of Samuel Johnson's conviction that otherwise there would be "confusion of progeny."

According to Engels, for example, the human family, developing "in the transitional period between the upper and middle stages of barbarism," was a harbinger of civilization. The family, he wrote,

> is based upon the supremacy of the man, the express purpose being to produce children of undisputed paternity; such paternity is demanded because these children are later to come into their father's property as his natural heirs. It is distinguished from pairing marriage [an earlier stage] by the much greater strength of the marriage tie, which can no longer be dissolved at either partner's wish. As a rule it is now only the man who can dissolve

it, and put away his wife. The right of conjugal infidelity also remains secured to him . . . and as social life develops he exercises
his right more and more; should the wife recall the old form of
sexual life and attempt to revive it, she is punished more severely
than ever.[38]

These issues surfaced again in the 1960s as part of feminist
doctrine. According to Mary Jane Sherfey, "One of the requi-
site cornerstones upon which all modern civilizations were
founded was *coercive* suppression of woman's inordinate sexuality." This theme is summarized well by Nancy Marval,
though I wonder whether she would not be a bit surprised to
know that several sociobiologists, approaching the problem
from a different angle, and (I suspect) completely independently, have ended up by concurring with her point by point.
Marval writes:

> In a patriarchal culture . . . sexuality is a crucial issue . . . men
> have no direct access to reproduction and survival of the species.
> As individuals, their claim to any particular child can never be as
> clear as that of the mother who demonstrably gave birth to that
> child . . . The only way a man can be absolutely sure that he is
> the one to have contributed that sperm is to control the sexuality
> of the woman . . .
> He may keep her separate from any other man as in a harem,
> he may devise a mechanical method of preventing intercourse
> like a chastity belt, he may remove her clitoris to decrease her
> erotic impulses, *or* he may convince her that sex is the same
> thing as love and if she has sexual relations with anyone else she
> is violating the sacred ethics of love.[39]

The only major difference between Marval's statement and
those of some sociobiologists[40] would be identification of the
enemy. Marval suggests it is "he" who is responsible for suppression of females. Sociobiologists, extending her arguments
to creatures other than *Homo sapiens,* point out that natural
selection—and not "men" *per se*—is responsible, and note
that females as well as males are implicated.

Lest anyone doubt that women collude in the supposed
"plot" to suppress them, consider the court ladies of ancient
China, the *nu shih* whose job it was to supervise the wives
and concubines of the emperor. As have mothers, fathers, in-
laws of both sexes, neighbors, and nosy parkers through the

ages, these women busied themselves with the sexual status and conduct of women—in this case, residents of the emperor's seraglio. Seldom has the scrutiny of women's sexual lives been raised to such a high level of professionalism. Modern scientists might well covet the detailed information available to the *nu shih*.

Elaborate protocol surrounded the hundreds of women housed in the harems of T'ang dynasty emperors; the ever increasing numbers made meticulous bookkeeping essential to the emperor's confidence in his paternity and to ensure that only his progeny received the benefits of imperial investment. In her records the *nu shih* entered the "date and hour of every successful sexual union, the dates of menstruation of every woman, and the first signs of pregnancy." In some cases, even more exacting measures were deemed necessary. The "Notes of the Dressing Room" by Chang Pi, a scholar whose teachings flourished around A.D. 940, records that at the beginning of the K'ai-yuan era (713–741) every woman with whom the emperor had slept was stamped on her arm and the mark then rubbed with cinnamon to make it indelible.[41] In earlier times, and perhaps also as late as the T'ang, as another form of recordkeeping, a silver ring was transferred from the woman's right leg to her left to mark each sexual union with the emperor. If she conceived, she was given a golden ring to wear.

Whole chapters of human history could be read as an effort to contain the promiscuity of women and thus to establish, from circumstantial evidence, the paternity that could never be proved directly (before the advent of sophisticated laboratories). Whatever the biological component may be, the behavioral component is readily demonstrated.

Only an anthropologist could have undertaken this task with the richness of ethnographic detail that has characterized the work of Mildred Dickemann. In a brilliant series of three papers, she has reviewed human practices that serve to cloister women and expropriate their fecundity.[42] She has focused particularly on stratified human societies, in which women typically marry "up" into families of higher standing than their own. This practice, called "hypergyny" by anthropologists, is common throughout much of the world. Dickemann has drawn much of her material from accounts of ancient China, medieval Europe, and north India just after its colonization by the British. She takes as given that higher status improves the reproductive success of men and women alike. Quantitative data in support of this point are hard to come by, but she points to the relationship throughout history between possession of property and the survival of family lines. She also cites the extreme vulnerability of the dispossessed, particularly during times of famine.

Dickemann's focus on stratified societies is important to her argument because it allows her to assume that a properly hypergynous marriage benefits not only the bride but her family. Her parents can look forward to grandchildren born into a world of improved opportunities; large dowries are understandable as the price paid for high-status grandsons, whose prospects include good health, longevity, perhaps multiple wives, and, crucially, many children of their own. The bride's family, then, as well as the groom's, has an interest in ensuring her virginity and future fidelity. Access through marriage or concubinage to a wealthy family is competitive, and the bride's family has a direct stake in her reputation and eligibility. Where these are connected to extreme standards of modesty, such as veiling women in purdah or strict seclusion,

Women in purdah

the bride's family forces her compliance, though in many cases special pressure is unnecessary because, as Dickemann points out, the woman herself is indoctrinated from infancy to value and manifest ideals of feminine modesty. Even without seraglios, the behavior of women is shaped through physical coercion (beatings or the threat of them) or, more commonly, superstition (the threat of "damnation" or rape by demons). In contemporary Maya-speaking communities throughout Central America, for example, female sexuality is considered extremely dangerous. Women thought to be careless or excessive in their sexuality—and "excessive" here includes any adoption by women of a copulatory position likely to increase clitoral stimulation—are publicly mocked or threatened with rape by a mythical demon so potent that the offending woman will give birth every night, night after night, until she swells up and dies.[43]

The more clearly stratified the society, and the higher the sights set by the family for a young girl's future, the easier it is to justify the time and expense (particularly in lost labor) of elaborate claustration. As Dickemann puts it, "Women's modesty and the investment of energy into its maintenance become marks of family pride, major indices of public reputation." By way of example, she cites the north Indian ethnographer Elizabeth Cooper:

> You can tell the degree of a family's aristocracy by the height of the windows in the home. The higher the rank, the smaller and higher are the windows and the more secluded the women. An ordinary lady may walk in the garden and hear the birds sing and see the flowers. The higher grade lady may only look at them from her windows, and if she is a very great lady indeed, this even is forbidden to her, as the windows are high up near the ceiling, merely slits in the wall for the lighting and ventilation of the room.[44]

At this higher grade of the social scale, it was common practice in north India to destroy daughters due to the impossibility for them to marry up and thereby improve the fortunes of the family. Better to invest entirely in sons and recruit wives from the lower ranks. Infanticide is widespread in traditional societies.[45] Unlike other animals, which are seldom inclined to kill their own young but rather only those belonging to competitors, humans dispatch their own infants for a variety of reasons, often due to economic constraints or inopportune timing of a birth, without regard to the infant's sex. But when a discrimination between offspring is made on the basis of sex, it is virtually always the female infant that is killed. Preferential female infanticide occurs in both stratified societies such as the Rajputs and in more egalitarian societies such as the Yanomamo Indians of South America, where sons are greatly valued as warriors and where, if women are in short supply, wives are obtained by raiding women from other villages.

Influential members of cultures throughout north India, ancient China, medieval Europe, and the Arab world (to name only some of the best-documented cases) have subscribed to a "Sherfeyian" assessment of women's readiness to engage in sexual activity. Whatever the biological facts, female sexuality has been seen as a force sufficiently real as to be worth guarding against. The very sexual feelings that are often interpreted as bonding a woman to her mate can also be viewed as forces which would incite a woman to extramarital activity.

Myriad rationalizations are offered for forbidding women their freedom of movement, for sequestering them. Some Muslims and Rajputs that I know argue that claustration of women is undertaken for their own good, to protect them from kidnap or rape. No doubt it does. The feet of daughters in ancient China might have been bound not to incapacitate them, to tether them forever on tiptoes, but to show off the lady's status and advertise the fact that her family could afford to do without her labor. Killing daughters might be explained away as simply a logical response to poor marriage possibilities in societies where a spinster's life is not worth living. One need not accept the interpretation that in hypergynous mating systems the inclusive fitness of the entire family is diminished if the resources of a high-ranking family are diverted to a daughter. Yet taken together, the evidence falls again and again into a similar configuration, making Mildred

Dickemann's sociobiological analysis of such practices a compelling one.

Furthermore, some practices, such as clitoridectomy, have no convincing alternative explanation and make sense in no other context. Efforts to explain away this practice by calling it "female circumcision"—a companion practice to removing the male foreskin as an initiation rite—are too obviously exercises in anthropological euphemism. They don't wash. Cultural beliefs can only be an overlay upon this straightforward effort to alter female anatomy through surgical removal of the clitoris or sewing together the lips of the vagina (infibulation). The consequences to male and female of the respective procedures are radically dissimilar. Male circumcision has little obvious consequence to sexuality. Clitoridectomy is an effective means of reducing sexual pleasure.

The earliest evidence for clitoridectomy dates from ancient Egypt, where both clitoral excision and labial fusions were practiced, as can be seen from female mummies. The Greek historian and geographer Agatharchides, who visited Ethiopia in the second century B.C., noted that the people there excised their women in the Egyptian tradition.[46] Perhaps the most remarkable aspect of female circumcision is that it continues to be practiced; one estimate holds that upward of 20 million women in the world today are affected,[47] despite the fact that it can cause serious medical complications, including septicemia, hemorrhage, and shock, as well as serious urinary and obstetrical problems and threats to female fertility. So extensive is the infibulation operation—the clitoris is excised and the surrounding tissue scarified so that the fusion of the labia will occur during healing—that approximately 9 percent of girls operated on under semimodern conditions (with some anesthesia) suffer hemorrhage or shock. Infibulated women are partially cut open at marriage, and must be fully opened at childbirth—after which they are sewn up again. Hence, the possibility of impaired fertility or death from the procedure persists throughout the woman's reproductive years. In a recent study of 4,024 women at Khartoum, in the Sudan, the percentage of infibulated women suffering from urinary infection was four times as great as noncircumcised women, the rate of chronic pelvic infection more than twice as high. Of 3,013 of these women who had been infibulated, 84 percent reported that they had never had an orgasm.[48] Far from finding intercourse pleasurable, many circumcised and infibulated women find it painful.

From a biological point of view, women can scarcely be said to benefit from this practice. Why, then, does the society subject women to an operation which both pains and endangers them? Why do men in some societies refuse to marry uncircumcised women? The obvious answer, recognized by feminists and sociobiologists alike, is that female circumcision increases the certainty of paternity and reduces the likelihood that a man will be cuckolded and thus tricked into supporting some other man's offspring. In fact, a commonly cited folk rationale for female circumcision is that the operation promotes chastity by reducing sexual desire.[49]

T HERE is a certain irony that any feminist should ever have undertaken intellectual lobbying to exclude knowledge about other nonhuman primates from efforts to understand the human condition. No doubt there seemed good reason at the time. Primatologists, and those social scientists who relied upon their work, made much of the supposedly greater competitive potential of males, the importance of males in structuring the society, and the apparent inability of females to maintain stable social systems.[50] Such models were too obviously a projection of androcentric fantasy. It was not unreasonable to fear that such views might lead to policies contrary to the aspirations of women. As late as 1980, one can find it seriously suggested that differences in competitiveness between girl and boy athletes may be "adaptive for our species" and therefore "should not be erased."[51] Yet if more reliable information about the actual workings of primate social systems had been available, quite a different portrait of females would have been painted. The real irony, though, is that women in so many human societies occupy a position that is far worse than that of females in all but a few species of non-human primates.

Incontestably, weaker individuals are often victimized by stronger ones. This can certainly be documented throughout the primates, but never on the scale with which it occurs among people, and never directed exclusively against a particular sex after the fashion of female infanticide, claustration of daughters and wives, infibulation, or the suttee—the immolation of widows which rather effectively preserves the chastity of the wife of a dead man. Only in human societies are females as a class systematically subjected to the sort of treatment that among other species would be rather randomly accorded to the more defenseless members of the group—the very young, the disabled, or the very old—regardless of sex. This is another way of saying that among people, the biological dimorphism of the sexes has become institutionalized. How did this come about?

Division of labor (and with it the potential for one person to benefit from another's work), the means to accumulate property and not just territory, and the organizational ability to allocate tasks—all of these altered a fundamental relationship between mates. Whereas among other mammals the most polygynous males are necessarily those which invested least in offspring, men could control both large numbers of women *and* the resources they needed to survive and reproduce.[52] Among other animals there is an inverse relationship between polygyny and male investment: males must sacrifice additional mates for the sake of rearing the offspring that they do sire by carrying, feeding, or protecting them. If high levels of male investment are required, male primates are monogamous. But this old relationship no longer necessarily holds in the human case. Polygyny coupled with substantial, albeit sometimes indirect, paternal investment is entirely possible. (*Vide* the three families of H. L. Hunt.) Polygyny can occur together with high levels of paternal investment and also—by virtue of either guards or gossips—surveillance.

The shift toward reckoning inheritance in the father's line and living near his birthplace has also had far-reaching consequences for women. Among all but a handful of other primates, females define a territory and occupy it from generation to generation. Only two of the three great apes and four apparently unusual species of monkeys ensure outbreeding through female transfer. But female transfer is the rule for most human societies. For women, the social consequences of joining the man's family have been well documented. The anthropologist Naomi Quinn sums up the straits of a bride moving in with her husband's people:

> Such a bride suffers the loneliness and the scrutiny of her affines which typifies the lot of all virilocally married women . . . ; in addition she may find herself under the authority of a hostile mother-in-law, whose interests are opposed to hers in competition for the affection and loyalty of her husband. Her only claim to status rests on her success in bearing and raising sons and her eventual position as a mother-in-law herself. Typically, women can only gain power in such households indirectly, through men, and their strategies for so doing may be characterized by gossip, persuasion, indirection and guile.[53]

Most importantly, except for those cases in which additional wives are typically close relatives of the first one (as in sororal polygyny), women exchanged in marriage with other villages are more or less cut off from the support of their relatives. Unrelated women marrying into the same families often have conflicts of interest which preclude either collective action or individual resistance.[54] For generations and generations, then, in populations where a woman moved away from her natal home, her freedom was severely constrained by the scrutiny of her husband's relatives.

Close observation of the sexual conduct of women is universal in human communities. Even in largely hunter-gatherer societies where property is not owned and women enjoy considerable freedom of action and movement—a freedom dictated by the community's reliance on food gathered by women—it is virtually impossible to keep sexual liaisons secret. As Lorna Marshall, who has recounted the lives of Kalahari desert foragers, puts it: "There is no privacy in a !Kung encampment, and the vast veld is not a cover. The very life of these people depends on their being trained from childhood to

look sharply at things . . . They register every person's foot-
prints in their minds . . . and read in the sand who walked
there and how long ago."[55] Worse, they talk about it among
themselves.

T HROUGHOUT much of evolutionary history, the uncertainty of paternity has been one of several advantages females retained in a game otherwise heavily weighted toward
male muscle mass. Female primates evolved a variety of strategies to pursue this advantage—the shift to situation-dependent receptivity, concealed ovulation, an assertive sexuality.
Such attributes improved the abilities of females to manipulate males and to elicit from them the care and tolerance
needed to rear the infants they bore. Females were abetted in
this by selection upon males themselves to promote the survival of infants which were likely, or even just possibly, their
own. But such advantages were not granted to females in a
vacuum; once again the ball was tossed back into the other
court. To keep women (and their sexuality) in check, husbands and their relations (and perhaps especially property-
owning families) devised cultural practices which emphasized
the subordination of women and which permitted males authority over them. Presumably, females have adapted to these
new constraints, becoming, among other things, more dis-
creet and more subdued, but the fact is that little is known
about the sexual life histories of women, and the matter of
their legacy will not be soon resolved.

The human world is radically different from that of other
primates. Human ingenuity, and with it the ability to build
walls, to count, to tell tales, to transport food and store it, and
particularly to allocate labor (to control not just the reproductive capacities but also the productive capacities of other individuals), all of these eroded age-old female advantages. Yet,
by the same token, in areas of the Near and Far East, in ancient Greece, in pockets of civilization in South America and
northern Europe, to take only those cases we have documentation for, the same ingenuity that permitted people to oversee
and manipulate complex transactions gave rise to standards of

morality which could be articulated and set down in the form of legal systems. In the Western world, the rights of "man" were gradually extended to both sexes. Women can now aspire to a degree of independence corresponding to that of men. In this respect people are in a class by themselves. Of all females, the potential for freedom and the chance to control their own destinies is greatest among women.

> [*First college president:*] *Dear Lord, what next. First blacks, now*
> *women.*
> [*Second college president:*] *Just give it another couple of years and it*
> *will all go back to normal.*
> Conversation in an elevator, overheard by a third college president,
> JAQUELYN MATTFELD OF BARNARD, 1979.

Afterword

This effort to correct a bias within evolutionary biology, to expand the concept of "human nature" to include both sexes,
continues an endeavor begun over a century ago. As early as
1875, Antoinette Brown Blackwell warned of the intellectual
hazard from imagining that only one sex evolved. A major goal
of this book has been to describe the female primates that did
evolve over the last seventy million years. By and large, these
females are highly competitive, socially involved, and sexually
assertive individuals. Competition among females is one of the
major determinants of primate social organization, and it has
contributed to the organisms women are today. Yet, social scientists have collected little information on this facet of feminine personalities. Never before now, I suspect, has it been so
important to take account of the full range of woman's nature.

Throughout millions of years of evolution, mammalian
mothers have differed from one another in two important
ways: in their capacity to produce and care for offspring and
in their ability to enlist the support of males, or at least to
forestall them from damaging their infants. Female primates
have differed from one another in their capacity to influence
the reproductive careers of their descendants. Here is a sex

wide open to natural selection, and evolution has weighed heavily upon it.

Those same forces which predisposed females to intelligence and assertiveness also selected the highly competitive individuals among them. This is the dark underside of the feminist dream. If it is shown—as I believe it will be—that there are no important differences between males and females in intelligence, initiative, or administrative and political capabilities, that women are no less qualified in these areas than men are, one has to accept also that these potentials did not appear gratuitously as a gift from Nature. Competition was the trial by fire from which these capacities emerged. The feminist ideal of a sex less egotistical, less competitive by nature, less interested in dominance, a sex that will lead us back to the "golden age of queendoms, when peace and justice prevailed on earth," is a dream that may not be well founded.

Widespread stereotypes devaluing the capacities and importance of women have not improved either their lot or that of human societies. But there is also little to be gained from countermyths that emphasize woman's natural innocence from lust for power, her cooperativeness and solidarity with other women. Such a female never evolved among the other primates. Even under those conditions most favorable to high status for females—monogamy and closely bonded "sisterhoods"—competition among females remains a fact of primate existence. In a number of cases it leads to oppression of some by others; in other cases competition among females has forestalled the emergence of equality with males. As it happens, a particular subset of human societies (patrilineal and stratified) takes the prize for "sexism." Yet the same human ingenuity that eroded the position of women in those cases scattered—in other soils—the seeds of sexual equality.

The female with "equal rights" never evolved; she was invented, and fought for consciously with intelligence, stubbornness, and courage. But the advances made by feminists rest on a precarious framework built upon a unique foundation of historical conditions, values, economic opportunities, heroism on the part of women who fought for suffrage, and perhaps especially technological developments which led to birth control and labor-saving devices and hence minimized physical differences between the sexes. This structure is fragile. Should it collapse, it is far from certain that the scaffolding needed to surmount oppressive natural and cultural barriers could ever be pieced together again.

To assume that women today are regaining a natural pre-eminence, or reinstating some original social equality, belittles the real accomplishment and underestimates its fragility. However well-intentioned, these myths pose grave dangers to the actual progress of women's rights. They devalue the unique advances made by women in the last few hundred years and tempt us to a false security. Injustices remain; there are abundant new problems; yet, never before—not in seventy million years—have females been so nearly free to pursue their own destinies. But it won't be easy.

A Simplified Classification of Living Primates

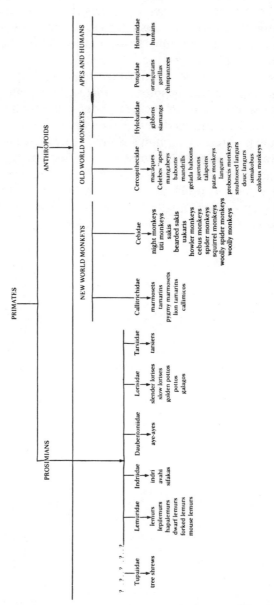

Adapted from *The Handbook of Living Primates*, by J. R. Napier and P. Napier (New York: Academic Press, 1967).

Taxonomy of the Primate Order

Suborder Prosimii	Prosimians
Infraorder Lemuriformes	Malagasy lemurs
Superfamily Lemuroidea	Lemuroids
Family Lemuridae	lemurs
Subfamily Lemurinae	
Genus *Lemur*	
Species *L. catta*	ringtailed lemur
L. variegatus	ruffed lemur
L. macaco	black lemur
L. mongoz	mongoose lemur
L. rubriventer	red-bellied lemur
L. fulvus	brown lemur
Hapalemur	gentle lemur
H. griseus	
H. simus	
Lepilemur	lepilemur
L. mustelinus	
Subfamily Cheirogaleinae	small nocturnal lemurs
Cheirogaleus	dwarf lemurs
C. major	greater dwarf lemur
C. medius	fat-tailed dwarf lemur
C. trichotis	hairyeared dwarf lemur
Microcebus	mouse lemur
M. murinus	
M. coquereli	Coquerel's mouse lemur
Phaner	forked lemur
P. furcifer	

Family Indriidae
    Indri   indri
      I. indri
    Avahi   woolly lemur, avahi
      A. laniger
    Propithecus   sifakas
      P. diadema   diademed sifaka
      P. verreauxi   white sifaka
Superfamily Daubentonioidea   aye-aye
  Family Daubentoniidae
    Daubentonia
      D. madagascariensis
Infraorder Lorisiformes   lorises, lorisiforms
  Family Lorisidae
    Subfamily Lorisinae
      Loris   slender loris
        L. tardigradus
      Nycticebus   slow loris
        N. coucang
      Arctocebus   golden potto
        A. calabarensis
      Perodicticus   potto
        P. potto
    Subfamily Galaginae   galagos, bushbabies
      Galago
    Subgenus *Galago*
        G. senegalensis   Senegal or lesser bushbaby
        G. crassicaudatus   thicktailed or greater bushbaby
        G. alleni   Allen's bushbaby
      Euoticus   needleclawed bushbaby
        G. elegantulus
        G. inustus
      Galagoides   dwarf bushbaby
        G. demidovii
Infraorder Tarsiiformes   tarsiers
  Family Tarsiidae
    Tarsius
      T. spectrum   spectral tarsier
      T. bancanus   Horsfield's tarsier
      T. syrichta   Philippine tarsier

Suborder Anthropoidea   monkeys and apes
  Superfamily Ceboidea   New World monkeys
    Family Callitrichidae   tamarins and marmosets
      Subfamily Callitrichinae
        Callithrix   marmosets
          C. jacchus   common marmoset
          C. argentata   blacktailed marmoset
          C. aurita   white-eared marmoset
          C. flaviceps   buffheaded marmoset

C. geoffroyi	whitefronted marmoset
C. penicillata	blackeared marmoset
C. humeralifer	Santarem marmoset
C. chrysoleuca	golden marmoset
Cebuella	pygmy marmoset
C. pygmaea	
Saguinus	hairy-faced tamarins
Saguinus	
S. tamarin	negro tamarin
S. devillei	De Ville's tamarin
S. fuscicollis	saddle back tamarin
S. fuscus	
S. graellsi	Rio Napo tamarin
S. illigeri	redmantled tamarin
S. imperator	emperor tamarin
S. melanoleucus	white tamarin
S. midas	black-faced tamarin
S. labiatus	redbellied tamarin
S. mystax	mustached tamarin
S. pileatus	redcapped tamarin
S. pluto	Lönnberg's tamarin
S. weddelli	Weddell's tamarin
S. nigricollis	black-and-red tamarin
S. lagonotus	harelipped tamarin
Oedipomidas	crested barefaced tamarins, pinchés
S. oedipus	pinché, cottontop
S. geoffroyi	Geoffroy's tamarin
Marikina	true barefaced tamarins
S. bicolor	pied tamarin
S. martinsi	Martin's tamarin
S. leucopus	whitefooted tamarin
S. inustus	
Leontideus	lion tamarins
L. rosalia	golden lion tamarins
L. chrysomelas	
L. chrysopygus	
Subfamily Callimiconinae	Goeldi's marmoset, callimico
Callimico	
C. goeldii	
Family Cebidae	
Subfamily Aotinae	
Aotus	night monkey, owl monkey
A. trivirgatus	
Callicebus	titis
C. personatus	masked titi
C. moloch	dusty titi, moloch
C. torquatus	widow monkey
Subfamily Pitheciinae	
Pithecia	sakis
P. pithecia	paleheaded saki
P. monachus	monk saki

Chiropotes	bearded saki
C. *satanas*	black saki
C. *albinasus*	whitenosed saki
Cacajao	uakaris
C. *melanocephalus*	blackheaded uakari
C. *calvus*	bald uakari
C. *rubicundus*	red uakari
Subfamily Alouattinae	howler monkeys
Alouatta	
A. *belzebul*	redhanded howler
A. *villosa*	mantled howler
A. *seniculus*	red howler
A. *caraya*	black howler
A. *fusca*	brown howler
Subfamily Cebinae	
Cebus	capuchin monkey, cebus
C. *capucinus*	whitethroated capuchin
C. *albifrons*	whitefronted capuchin
C. *nigrivittatus*	weeper capuchin
C. *apella*	blackcapped capuchin
Saimiri	squirrel monkey
S. *sciureus*	squirrel monkey
S. *oerstedii*	redbacked squirrel monkey
Subfamily Atelinae	
Ateles	spider monkey
A. *paniscus*	black spider monkey
A. *belzebuth*	longhaired spider monkey
A. *fusciceps*	brownheaded spider monkey
A. *geoffroyi*	blackhanded spider monkey
Brachyteles	woolly spider monkey
B. *arachnoides*	
Lagothrix	woolly monkey
L. *lagotricha*	Humboldt's woolly monkey
L. *flavicauda*	Hendee's woolly monkey
Superfamily Cercopithecoidea	Old World monkeys
Family Cercopithecidae	
Subfamily Cercopithecinae	
Macaca	macaques
M. *sylvanus*	barbary macaque
M. *sinica*	toque macaque
M. *radiata*	bonnet macaque
M. *silenus*	liontailed macaque
M. *nemestrina*	pigtailed macaque
M. *fascicularis*	crabeating macaque
M. *mulatta*	rhesus macaque
M. *assamensis*	Assamese macaque
M. *cyclopis*	Formosan rock macaque
M. *arctoides*	stumptailed macaque
M. *fuscata*	Japanese macaque
M. *maurus*	Celebes or moor macaque
M. *thibetana*	Thibetan macaque

Cynopithecus	Celebes black ape
C. niger	
Cercocebus	mangabeys
C. albigena	graycheeked mangabey
C. aterrimus	black mangabey
C. torquatus	whitecollared mangabey
C. atys	sooty mangabey
C. galeritus	agile mangabey
Papio	baboons
P. cynocephalus	yellow baboon
P. anubis	olive baboon
P. papio	Guinea baboon } savanna baboons
P. ursinus	chacma baboon
P. hamadryas	sacred baboon, hamadryas
Mandrillus	mandrills
M. sphinx	drill
M. leucophaeus	mandrill
Theropithecus	gelada baboon
T. gelada	
Cercopithecus	guenons
Cercopithecus	
C. aethiops	vervet
C. sabaeus	green monkey
C. cephus	mustached monkey
C. diana	diana monkey
C. lhoesti	L'Hoest's monkey
C. preussi	Preuss's monkey
C. hamlyni	Hamlin's or owlfaced monkey
C. mitis	blue monkey
C. albogularis	Sykes's monkey
C. mona	mona monkey
C. campbelli	Lowe's guenon
C. wolfi	Wolf's monkey
C. denti	Dent's monkey
C. pogonias	crowned guenon
C. neglectus	De Brazza's monkey
C. nictitans	spotnosed monkey
C. petaurista	lesser spotnosed monkey
C. ascanius	redtail
C. erythrotis	redeared guenon
C. erythrogaster	redbellied guenon
Miopithecus	
C. talapoin	talapoin, dwarf guenon
Allenopithecus	
C. nigroviridis	Allen's swamp monkey
Erythrocebus	patas
E. patas	
Subfamily Colobinae	
Presbytis	langurs
P. aygula	Sunda Island langur
P. melalophos	banded leaf monkey

P. frontatus	whitefronted leaf monkey
P. rubicundus	maroon leaf monkey
P. entellus	Hanuman langur
P. senex	purplefaced leaf monkey
P. johnii	nilgiri langur
P. cristata	silvered leaf monkey, lutong
P. pileatus	capped langur
P. geei	golden langur
P. obscurus	dusky leaf monkey
P. phayrei	Phayre's leaf monkey
P. francoisi	François' leaf monkey
P. potenziani	Mentawei Island monkey
Rhinopithecus	snubnosed langur
R. roxellanae	golden monkey
R. avunculus	Tonkin snubnosed monkey
Pygathrix	douc langur
P. nemaeus	
Nasalis	proboscis monkey
N. larvatus	
N. concolor	simakobu
Colobus	colobus, guerezas
Colobus	
C. polykomos	King colobus ⎫ black-and-white
C. guereza	Abyssinian colobus ⎭ colobus monkeys
Procolobus	
C. verus	olive colobus
Piliocolobus	
C. badius	*red colobus*
C. kirkii	Kirk's colobus
Superfamily Hominoidea	Apes and humans
Family Hylobatidae	
Hylobates	gibbons
H. lar	white-handed or lar gibbon
H. moloch	silvery gibbon or moloch
H. pileatus	pileated gibbon
H. muelleri	Müller's gibbon
H. agilis	dark-handed or agile gibbon
H. hoolock	hoolock gibbon
H. concolor	concolor gibbon
H. klossi	Kloss's gibbon
Symphalangus	siamang
S. syndactylus	
Family Pongidae	great apes
Pongo	orangutan
P. pygmaeus	
Pan	chimpanzees
P. troglodytes	chimpanzee
P. paniscus	pygmy chimpanzee, bonobo
Gorilla	gorilla

G. *gorilla*
Family Hominidae
    Homo   humans
    H. *sapiens*

This taxonomy is adapted from the *Handbook of Living Primates* by J. and P. Napier (New York: Academic Press, 1967) and from *The Evolution of Primate Behavior* by Alison Jolly (New York: Macmillan, 1972). Minor changes have been made to conform to current usage (for example, *Presbytis cristatus* is now known as *Presbytis cristata*). I have added a lemur (*Lemur fulvus*) and, following David Chivers' 1977 article "The lesser apes" (cited in the text), two additional gibbon species (*Hylobates pileatus* and *H. muelleri*). I have left the Callitrichidae as Alison Jolly cites them, but readers should be aware that some taxonomists prefer to lump together species of tamarins and marmosets which are listed here as separate. Finally, in a number of places I have altered "common" names to conform with the name used by the researchers whose work I cite in the text. Hence I call the pagai island langur (*Nasalis concolor*) by its native name, simakobu; the variegated lemur (*Lemur variegatus*) has become the ruffed lemur; Campbell's monkey (*Cercopithecus campbelli*) is Lowe's guenon; and so on.

This is a conventional taxonomy, but it is not engraved in stone. After centuries of classifying primates, experts still disagree.

Notes

1. Some Women That Never Evolved

1. Nancy Chodorow's otherwise very apt essay, "Being and doing: a cross cultural examination of the socialization of males and females," in *Women in Sexist Society*, eds. Vivian Gornick and Barbara K. Moran (New York: New American Library, 1972), equates "biologically derived" with "inescapable." See also Barbara Chasin, "Sociobiology: a sexist synthesis," *Science for the People*, May–June 1977, and Freda Salzman, "Are sex roles biologically determined?" *Science for the People*, July–August 1977. According to Salzman, "In the past ten years, a succession of highly publicized works have purported to demonstrate that women's subordinate position in our society is due, in good part, to innate (genetic) differences between males and females, and not to external factors as claimed by the women's movement. These theories are simply new constructions of that old theme, 'Biology is Destiny.' Furthermore, these theories argue that sex roles are resistant to change and that dire consequences will result if we try to change them" (p. 27).

Because criticisms such as Salzman's have been directed rather specifically at the work of Edward O. Wilson, it seems worth juxtaposing here what his current position on sex roles actually is. The following quotation is from Wilson's 1978 essay, *On Human Nature* (Cambridge: Harvard University Press): "Here is what I believe the evidence shows—modest genetic differences exist between the sexes; the behavioral genes interact with virtually all existing environments to create a noticeable divergence in early psychological development; and the divergence is almost always widened in later psychological development by cultural sanctions and training. Societies can probably cancel the modest genetic differences entirely by careful planning and training . . . So at birth the twig is already bent a little bit—what are we to make of that? It suggests that the universal existence of sexual division of labor is not entirely an accident of cultural evolution. But it also supports the conventional view that the enormous variation among societies in the degree of that division is due to cultural evolution" (pp. 129–132).

2. Consider the following personality profile extracted from a 1963 essay entitled "The Female Primate": "Her primary focus . . . is motherhood . . . Her dominance status, associations with adult males, female companions, and daily activities to a large extent are a function of her status as a mother and her phase of the reproductive cycle . . . Dominance interaction is usually minimal in the life of a female. She is invariably subordinate to all the adult males in the group and seldom if ever contests his superior status . . . Sexual behavior actually plays a small part in the life of an adult female. . . . A female is sexually receptive . . . for from five to seven days a month when she is not pregnant . . . A female's dominance status is not constant. Instead, it changes as she enters different phases of her reproductive cycle." Phyllis Jay, "The female primate," in *The Potential of Woman*, eds. Seymour Farber and Roger Wilson (New York: McGraw-Hill, 1963), pp. 3–7.

3. Lionel Tiger, *Men in Groups* (New York: Vintage, 1970), p. 45.

4. Evelyn Reed, *Woman's Evolution: From Matriarchal Clan to Patriarchal Family* (New York: Pathfinder Press, 1975), p. 52.

5. Eleanor Emmons Maccoby and Carol Nagy Jacklin, *The Psychology of Sex Differences*, vol. 1 (Stanford: Stanford University Press, 1974), p. 256.

6. Lionel Tiger, "The possible biological origins of sexual discrimination," in *Biosocial Man*, ed. Don Brothwell (London: Eugenics Society, 1977), p. 28.

7. Even before much of the information reviewed in this book became available, several authors suspected that female dominance relations were exceedingly important for the evolution of primate societies. Jane Lancaster wrote a highly prescient article entitled "In praise of the achieving female monkey" which emphasized that female matrilines formed the core of many primate troops and that it was primarily females who transmitted knowledge between generations. Lancaster was following the lead here of the pioneering British primatologists Vernon Reynolds and Thelma Rowell. Both Reynolds and Rowell have long been saying that female–female coalitions would turn out to be very important. Lancaster's article appeared in

the September 1973 issue of *Psychology Today*. See also her more recent article "Sex roles in primate societies," in *Sex Differences*, ed. M. S. Teitelbaum (New York: Doubleday/Anchor Press, 1976). See also Thelma Rowell, *Social Behaviour of Monkeys* (Middlesex: Penguin Books, 1972); M. K. Martin and B. Voorhies, *Female of the Species* (New York: Columbia University Press, 1975).

8. See, for example, Naomi Weinstein's critique of primate studies in her article "Psychology constructs the female," in *Women in Sexist Society* (note 1).

9. Michelle Zimbalist Rosaldo and Louise Lamphere, "Editor's introduction," in *Woman, Culture and Society* (Stanford: Stanford University Press, 1974), p. 3. The conclusion that in virtually all human societies political power is wielded by men is not an idiosyncratic or isolated viewpoint. Rather, it represents a consensus among such highly respected women anthropologists as Kathleen Gough (cited by Rosaldo and Lamphere), M. K. Martin and B. Voorhies (see note 7), and Ernestine Friedl. Much of the controversy over male dominance revolves around the question of *inevitability*, rather than the issue of whether or not male dominance exists. In Friedl's words, "The environmentalists do not deny the existence of male dominance in all known societies. Their argument is that it is not inevitable and need not remain a permanent state of affairs," E. Friedl, *Women and Men: An Anthropologist's View* (New York: Holt, Rinehart and Winston, 1974). For an important dissenting opinion, however, see Alice Schlegel's introductory and concluding chapters in her edited volume *Sexual Stratification: A Cross-Cultural View* (New York: Columbia University Press, 1977). Schlegel stresses the variability of sex-related statuses in different societies. In her words: "Sexual stratification . . . is not panhuman but rather poses a problem that must be explained, for each society in terms of the forces to which it is responsive" (p. 356).

10. Eleanor Leacock's introduction to *The Origin of the Family, Private Property and the State* by F. Engels (New York: International Publishers, 1972; English translation of the 1884 German original). See also Karen Sacks, *Sisters and Wives: The Past and Future of Sexual Equality* (Westport, Conn.: Greenwood Press, 1979).

11. Nancy Chodorow, "Family structure and feminine personality," in *Woman, Culture and Society* (note 1). See also the expanded argument in Chodorow's recent book, *The Reproduction of Mothering* (Berkeley: University of California Press, 1978).

12. See for example the recent examination and refinement of this approach by Sherry Ortner, "Is female to male as nature is to culture?" in *Woman, Culture and Society* (note 1).

13. Sherwood Washburn and C. S. Lancaster, "The evolution of hunting," in *Man the Hunter*, eds. R. B. Lee and I. DeVore (Chicago: Aldine, 1968).

14. Tiger (note 3).

15. Claude Lévi-Strauss, *Les structures élémentaires de la parenté* (Paris: Plon, 1949).

16. Margaret Hennig and Anne Jardim, *The Managerial Woman* (New York: Pocket Books, 1977).

17. Sally Linton, "Woman the gatherer: male bias in anthropology," in *Toward an Anthropology of Women,* ed. Rayna Reiter (New York: Monthly Review Press, 1975), and Nancy Tanner and Adrienne Zihlman, "Woman in evolution, p. 1: Innovation and selection in human origins," *Signs* 1 (3):289–303 (1976). There is now an impressive literature pointing to the importance of woman's economic participation in determining her social status. "The most important clue to women's status anywhere," writes Ruby Leavitt, "is her degree of participation in economic life and her control over property and the products she produces, both of which factors appear to be related to the kinship system of a society." "Women in other cultures," in *Women in Sexist Society* (note 1). This importance of women's contribution to subsistence, and particularly their control of distribution of products outside the home, is examined by Friedl (note 9), and by Peggy Sanday, "Female status in the public domain," in *Woman, Culture and Society* (note 9). For an excellent general review of the literature in this area see Naomi Quinn, "Anthropological studies on women's status," *Annual Review of Anthropology* 6:181–225 (1977).

18. Seymour Parker and Hilda Parker, "The myth of male superiority: rise and demise," *American Anthropologist* 81 (2):289–309 (1979).

19. Mary-Claire King and A. C. Wilson, "Evolution at two levels in humans and chimpanzees," *Science* 188:107–116 (1975).

20. S. Sands and A. A. Wright, "Primate memory: retention of serial list items by a rhesus monkey," *Science* 209:938–940 (1980).

21. J. A. R. A. M. Van Hoof, "A comparative approach to the phylogeny of laughter and smiling," in *Non-Verbal Communication,* ed. R. A. Hinde (Cambridge: Cambridge University Press, 1972), pp. 209–241.

22. Martha McClintock, "Menstrual synchrony and suppression," *Nature* 229:244–245 (1971). Hans Kummer, *The Social Organization of Hamadryas Baboons* (Chicago: University of Chicago Press, 1968), pp. 176–177.

23. William McGrew, "Evolutionary implications of sex differences in chimpanzee predation and tool use," in *The Great Apes,* eds. D. Hamburg and E. McCown (Menlo Park: Benjamin/Cummins, 1979), pp. 441–463.

24. Jane B. Lancaster and Phillip Whitten, "Family matters," *The Sciences* 20(1):10–15 (1980).

25. Johan Bachofen, *Das Mutterrecht* (Basel: Beno Schwabe,

1961). See also, *Myth, Religion and Mother Right: Selected Writings of J. J. Bachofen*, trans. Ralph Manheim (Princeton: Princeton University Press, 1967).

26. Carolyn Fluehr-Lobban reviews the relevant literature in her timely "Marxist reappraisal of the Matriarchate," *Current Anthropology* 20(2):341–359 (1979). Fluehr-Lobban concludes that "there is no indication in the data so far compiled on matrilineality to support the contention that matriliny is a general stage in culture history." Furthermore, "where it is found, in horticultural societies, it appears to be a highly specific response to particular conditions in which the organized work of woman is a key to subsistence." There is even less evidence for a universal stage where women held power.

27. Charlotte Perkins Gilman, *Herland* (rpt. New York: Pantheon Books, 1979), p. 67. Long out of print, Gilman's novel was republished through the efforts of the historian Ann J. Lane, who also provides an introductory biography of Gilman's life.

28. Valerie Solanis, *The S.C.U.M. Manifesto* (New York: Olympic Press, 1967).

29. Elizabeth Gould Davis, *The First Sex* (New York: G. P. Putnam's Sons, 1971; rpt. Middlesex: Penguin Books, 1976).

30. Antoinette Brown Blackwell, *The Sexes throughout Nature* (New York: B. Putnam and Sons, 1875; rpt. Westport, Conn.: Hyperion Press, 1976), p. 16.

31. For a review of the relevant literature, see especially Richard Hofstadter, *Social Darwinism in American Thought* (Boston: Beacon Press, 1955). Social Darwinism continues to be an important force in popular thinking. Robert Burton's recent book, *The Mating Game* (New York: Crown Publishers, 1976), p. 155, provides an excellent case in point. He writes: "The double moral standard which punishes an adulteress severely while often condoning the man can be defended on biological grounds. It increases a man's reproductive potential and it might be added that those who indulge in extramarital activities are those who are the 'fittest' and most deserving to be biological fathers as they must possess a high degree of cunning and initiative, and often physical agility."

32. There has been a tendency to identify the field of sociobiology with a single monumental book, *Sociobiology: The New Synthesis* by Edward O. Wilson (Cambridge: Harvard University Press, 1975). Unquestionably, Wilson's book is the most clearly written and comprehensive description of the sociobiological approach, but, as Mildred Dickemann pointed out in a recent critique, there is no monolithic body of knowledge identical with the field of sociobiology. Rather, there is a wide divergence of opinion on scores of issues among those who would call themselves sociobiologists. See Dickemann's "Comment on van den Berghe's and Barash's sociobiology," *American Anthropologist* 81(2):351–357 (1979).

2. An Initial Inequality

1. Katherine Ralls, "Mammals in which females are larger than males," *Quarterly Review of Biology* 51:245–276 (1976).

2. Donald Symons, *The Evolution of Human Sexuality* (New York: Oxford University Press, 1979), p. 140. For similar views, see Richard D. Alexander, *Darwinism and Human Affairs* (Seattle: University of Washington Press, 1979), p. 159, or the sociologist Randall Collins, "A conflict theory of sexual stratification," *Social Problems* 19:3–20 (1971).

3. Alice Schlegel, "Towards a theory of sexual stratification," in *Sexual Stratification: A Cross-Cultural View* (New York: Columbia University Press, 1977), esp. p. 14.

4. Among langur monkeys, for example, an aggressive young female still short of full adult body weight can routinely displace and hence outrank in the dominance hierarchy an older female weighing some 5 to 10 pounds more than she does. Although most male "troopleaders" we have weighed are about 40 pounds (compared with 20 to 30 pounds for females), males weighing as little as 35 pounds have been able to usurp a troop and maintain control in the face of competition from much larger males. S. Blaffer Hrdy, *The Langurs of Abu* (Cambridge: Harvard University Press, 1977), and additional data collected by Dan and Sarah Hrdy and by James Moore between 1977 and 1980.

5. G. A. Parker, R. R. Baker, and V. G. F. Smith, "The origin and evolution of gamete dimorphism and the male-female phenomenon," *Journal of Theoretical Biology* 36:529–553 (1972). For a less technical account, see Richard Dawkins' beautifully written book *The Selfish Gene* (Oxford: Oxford University Press, 1976).

6. Some women biologists have objected to what they feel is an overemphasis on the theory that females invest more in the production of offspring than males do. Such objections are valid in that the theory of anisogamy has been used too deterministically and without adequate consideration of such important exceptions as those species of cockroaches and butterflies where male sperm is accompanied by various nutrients in addition to the genetic material, or the many birds and mammals where male care is essential to survival of offspring. Nevertheless, critics of the theory of anisogamy often suffer from an unrealistic assessment of just how difficult it is in the natural world to obtain the resources necessary to reproduce—what I call "an American supermarket mentality."

"Does it really take more 'energy' to generate the one or relatively few eggs than the large excess of sperms required to achieve fertilization?" asks one critic, the biologist Ruth Hubbard in "Have only men evolved" in *Women Look at Biology Looking at Women*, eds. R. Hubbard, M. Henifin, and B. Fried (Cambridge: Schenkman, 1979),

p. 25. At their most extreme, I have heard such critics downplay as "trivial" the energetic costs of egg production, gestation, and even lactation. Such criticisms assume an infinite resource base and are not pertinent to mammalian evolution.

7. G. C. Williams, *Sex and Evolution* (Princeton: Princeton University Press, 1975). J. Maynard Smith, *The Evolution of Sex* (Cambridge: Cambridge University Press, 1978). See also W. D. Hamilton's review of Williams' book, appropriately and poetically entitled "Gamblers since life began: barnacles, aphids, elms," *Quarterly Review of Biology* 50:175–180 (1975).

8. Robert L. Trivers, "Parental investment and sexual selection," in *Sexual Selection and the Descent of Man*, ed. B. Campbell (Chicago: Aldine, 1972). Trivers in turn was influenced by G. C. Williams' *Adaptation and Natural Selection* (Princeton: Princeton University Press, 1966), and by an early article by A. J. Bateman, "Intrasexual selection in Drosophila," *Heredity* 2:349–368 (1948).

9. T. Gill, "The eared seal," *American Naturalist* 4(11):675–684 (1871). A century earlier Jean-Jacques Rousseau remarked upon the association between monogamy and reduced male–male competition: "Among the monogamous species, where intercourse seems to give rise to some sort of moral bond, a kind of marriage, . . . the male . . . is less uneasy at the sight of other males and lives more peaceably with them. Among these species the male shares the care of the little ones." *Émile* (New York: Dutton, Everyman's Library, 1974), p. 393. Three recent articles support these (particularly Gill's) observations: R. D. Alexander, L. Hoogland, R. D. Howard, K. M. Noonan, and P. W. Sherman, "Sexual dimorphisms and breeding systems in pinnipeds, ungulates, primates and humans," in *Evolutionary Biology and Human Social Organization* (North Scituate, Mass.: Duxbury Press, 1979); A. Gautier-Hion, "Dimorphisme sexuel et organisation sociale chez les cerocopithecines forestiers africains," *Mammalia* 39:365–374 (1975); T. Clutton-Brock, P. Harvey, and B. Rudder, "Sexual dimorphism, socionomic sex ratio and body weight in primates," *Nature* 269:797–800 (1977).

10. P. S. Rodman, "Individual activity profiles and the solitary nature of orang-utans," in *Perspectives on Human Evolution*, vol. 4: *The Behavior of the Great Apes*, eds. D. L. Hamburg and J. Goodall (London: W. A. Benjamin, 1979).

11. R. D. Martin, "Reproduction and ontogeny in tree shrews (*Tupaia belangeri*) with reference to their general behavior and taxonomic relationships," *Zeitschrift für Tierpsychologie* 25(4):409–495; 25(5):505–532 (1968).

12. James C. Hyatt, "Costs of being a parent keep going higher," *Wall Street Journal*, October 2, 1980, p. 33.

13. R. H. MacArthur and E. O. Wilson, *The Theory of Island Biogeography* (Princeton: Princeton University Press, 1967).

14. J. van Lawick Goodall, "Cultural elements in a chimpanzee community," *Symposia of the Fourth International Congress of Primatology* (Basel: S. Karger, 1973), pp. 144–184.

15. Jeanne Altmann, *Baboon Mothers and Infants* (Cambridge: Harvard University Press, 1980).

16. Caroline Pond, "The significance of lactation in the evolution of mammals," *Evolution* 31:177–199 (1977).

17. A. M. Thomson, "Maternal stature and reproductive efficiency," *Eugenics Review* 51:157–162 (1959), cited in Ralls (note 1).

18. Hans Kruuk, *The Spotted Hyena* (Chicago: University of Chicago Press, 1972). Kruuk originally hypothesized that the elaborate genitalia of the female hyena evolved to facilitate recognition in greeting ceremonies. However, recent authors favor the hypothesis presented here, that the peniforme clitoris and false scrotum in female hyenas is a by-product of high levels of androgens circulating in the fetal bloodstream. These authors compare the hyena case with that of baby girls whose pregnant mothers unwittingly took an androgen-like drug which prevented miscarriage but which also had the unfortunate side-effect of causing baby girls to be born with masculinized genitalia. See Stephen Gould, "Hyena myths and realities," *Natural History* 90(2):16–24 (1981), and P. A. Racey and J. D. Skinner, "Endocrine aspects of sexual mimicry in spotted hyenas, *Crocuta crocuta,*" *Journal of Zoology* 187:315–326 (1979).

3. Monogamous Primates: A Special Case

1. Devra Kleiman, "Monogamy among mammals," *Quarterly Review of Biology* 52:36–69 (1976).

2. Little is known about tree shrews and tarsiers in the wild. Their inclusion here among monogamous primates is based upon the very recent fieldwork of the Kawamichis with tree shrews (*Tupaia glis*) in Singapore, Malaysia; of John MacKinnon on the spectral tarsier (*Tarsius spectrum*) in North Sulawesi, in the Celebes; and of Carsten Niemitz on Horsfield's tarsier (*T. Bancanus*) in Sarawak, Borneo. T. Kawamichi and M. Kawamichi, "Social organization of tree shrew (*Tupaia glis*)," paper presented at the Eighth International Congress of Primatology, Florence, Italy, July 1980; J. MacKinnon, "The spectral tarsier—behavior in the wild and phylogenetic implications," paper presented at the Seventh International Congress of Primatology, Bangalore, India, January 1979; and C. Niemitz, "Outline of the behavior of *Tarsius bancanus,*" in *The Study of Prosimian Behavior,* eds. G. Doyle and R. D. Martin (New York: Academic Press, 1979).

Prosimians generally are not as well studied as the monkeys and apes, and detailed observations have been undertaken for only a few species. However, on the basis of rather limited census information, 5 additional Madagascar prosimians might be added to the 37 species listed in Table 1: *Avahi laniger; Lemur rubriventer; Varecia variegata; Hapalemur griseus;* and *Propithecus diadema.* These 5 species are consistently reported in small, apparently "family" groups. J. I. Pollock, "Spatial distribution and ranging behavior in lemurs," in *The Study of Prosimian Behavior* (above).

3. In a recent comprehensive review of "The evolution of monogamy: hypotheses and evidence," *Annual Review of Ecology and Systematics* 11:197–232 (1980). James F. Wittenberger and Ronald L. Tilson discuss five hypotheses to explain monogamy: (1) Monogamy should evolve when male parental care is nonshareable and indispensable to female reproductive success. (2) Monogamy should evolve in territorial species if a female benefits more from mating with an unmated male than with a male who is already mated. (3) Monogamy should evolve in nonterritorial species when the majority of males can reproduce most successfully by defending exclusive access to a single female. (4) Monogamy should evolve if aggression by females prevents a male from acquiring two mates. (5) Monogamy should evolve when males are less successful with two mates than with one. According to Wittenberger and Tilson, hypotheses 2, 3, and 4 were sufficient to explain all cases of monogamy in mammals.

4. A detailed account of food sharing among marmosets is provided by K. Brown and D. S. Mack in "Food-sharing among captive *Leontopithecus rosalia,*" *Folia Primatologica* 29:268–290 (1978). Food sharing has also been described for wild titi monkeys by Dawn Starin in "Food transfer by wild titi monkeys (*Callicebus torquatus torquatus*)," *Folia Primatologica* 30:145–151 (1978), and for captive gibbons by Teryl Schessler and Leanne Nash in "Food sharing among captive gibbons (*Hylobates lar*)," *Primates* 18(3):677–689 (1977). Occasional food sharing has also been described among polygynous species such as captive spider monkeys, stump-tailed macaques, savanna baboons, and douc langurs, as well as wild chimpanzees. As described by W. C. McGrew, "Patterns of plant food sharing by wild chimpanzees," in *Contemporary Primatology* (Basel: S. Karger, 1975), shared food among *Pan troglodytes* typically involves the transfer of vegetable foods from mother to offspring except in the case of meat, which may be distributed among adult males, a process described in detail by G. Teleki, *The Predatory Behavior of Wild Chimpanzees* (Lewisburg, Pa.: The Bucknell University Press, 1973). According to Richard Wrangham, however, male chimpanzees *only* give up meat when satiated, or to reduce competition for the meat sufficiently to permit the initial possessor to consume the prey without harassment; see Wrangham's "The Be-

havioral ecology of chimpanzees in Gombe National Park, Tanzania," Ph.D. thesis presented to Cambridge University, 1975. Such sharing is probably not comparable to the voluntary provisioning of infants by males in monogamous species. More tantalizing is a single account of food sharing in the little-known pygmy chimpanzee, *Pan paniscus.* As in *Pan troglodytes,* mothers share food with offspring, but on one occasion a dominant male in the group was also seen to provide an infant with a fruit. Suehisa Kuroda, "Social behavior of the pygmy chimpanzees," *Primates* 21(2):181–197 (1980).

5. W. Leutenegger, "Maternal–fetal weight relationships in primates," *Folia Primatologica* 20:280–293 (1973). See also Susan M. Ford, "Callitrichids as phyletic dwarfs, and the place of the Callitrichidae in Platyrrhini," *Primates* 21(1):31–43 (1980).

6. J. F. Eisenberg, "Comparative ecology and reproduction of New World monkeys," in *The Biology and Conservation of the Callitrichidae,* ed. D. Kleiman (Washington: Smithsonian Institution, 1978). Note that the terms *r-* and *K-*selected are relative concepts; they may be used to distinguish between groups (for example, chimpanzees are more *K-*selected than marmosets) but also to describe adaptive shifts within the same species or group along a continuum of differing environmental conditions. For example, as their habitat becomes densely populated, marmosets may produce fewer young each year, shifting toward a more *K-*selected strategy. Nevertheless, even these marmosets are more *r-*selected than any chimp could ever be. Fuzzy and imprecise as they are, the concepts of *r-* and *K-*selection provide a widely employed shorthand for describing natural populations.

7. George Edwards, *Gleanings of Natural History,* vol. 5 (London: College of Physicians, 1758), pl. 218.

8. R. J. Hoage, "Parental care in *Leontopithecus rosalia rosalia*: sex and age differences in carrying behavior and the role of prior experience," in *Biology and Conservation of the Callitrichidae* (note 6).

9. Gisela Epple summarizes current research on the social systems of captive and wild marmosets in "The behavior of marmoset monkeys (Callitrichidae)," in *Primate Behavior,* vol. 4, ed. L. Rosenblum (New York: Academic Press, 1975), and in her recent update of that article, "Reproductive and social behavior of marmosets with special reference to captive breeding," in *Primates in Medicine,* eds. E. I. Goldsmith and J. Moor-Jankowski (Basel: S. Karger, 1978).

10. The recent field studies of *Saguinus oedipus* by P. F. Neyman in Colombia and by Gary Dawson in Panama are reported in *Biology and Conservation of the Callitrichidae* (note 6). Dawson's

findings can be consulted in greater depth in his 1976 Ph.D. thesis, "Behavioral ecology of the Panamanian tamarin, *Saguinus oedipus* (Callitrichidae, Primates)," presented to Michigan State University.

11. J. P. Hearne, "The endocrinology and reproduction in the common marmoset *Callithrix jacchus*," in *Biology and Conservation of the Callitrichidae* (note 6). Experimental studies by Y. Katz and G. Epple with saddle-back tamarins showed that when adult daughters were removed from their families and paired with an adult male, there was a rapid increase of estradiol levels in their urine and an onset of regular cyclicity in previously acyclic females. "Social influences on urinary estradiol cyclicity of female *Saquinus fuscicollis* (Callitrichidae)," *Antropologica Contemporanea* 3(2):219 (abstract) (1980).

12. Hugo van Lawick, *Solo: The Story of an African Wild Dog Puppy and Her Pack* (London: Collins, 1973). Jane Goodall, "Infant killing and cannibalism in free-living chimpanzees," *Folia Primatologica* 28:259–282 (1977).

13. Ronald L. Tilson, "Social organization of simakobu monkeys (*Nasalis concolor*) in Siberut Island, Indonesia," *Journal of Mammalogy* 58:202–212 (1977).

14. R. L. Tilson and R. R. Tenaza, "Monogamy and duetting in an Old World monkey," *Nature* 263:320–321 (1976); and A. Gautier-Hion and J. P. Gautier, "Le singe de Brazza: une stratégie originale," *Zeitschrift für Tierpsychologie* 46:84–104 (1978).

15. The skins of museum specimens are far from ideal as estimates of body length, but in the case of a species so rare and so endangered as *Presbytis potenziani*, better information may not be forthcoming. With this in mind, Eric Delson and I compared the skin lengths of male and female specimens from the American Museum of Natural History in New York: whereas the average male was much longer than the female among *Nasalis concolor* (9 adult males averaged 563.77 cm; 7 females averaged 533.28 cm), among *Presbytis potenziani* there was little difference in length (8 adult males averaged 557.75 cm and 9 females averaged 557.22 cm).

16. R. R. Tenaza, "Songs, choruses and countersinging of Kloss' gibbons (*Hylobates klossii*) in Siberut Islands, Indonesia," *Zeitschrift für Tierpsychologie* 40:37–52 (1976).

17. J. Marshall and E. Marshall, "Gibbons and their territorial songs," *Science* 193:235–237 (1976).

18. D. J. Chivers, "The lesser apes," in *Primate Conservation,* eds. H. S. H. Prince Rainier and G. Bourne (New York: Academic Press, 1977).

19. In addition to the three dichromatic gibbons, there are four other primates in which males and females are colored differently: the saki (*Pithecia pithecia*); a population of *Lemur mongoz* on the

Comoro Islands; *Lemur macacao;* and *Alouatta caraya,* a species of howler monkey living in Argentina. It has recently been suggested that *Alouatta fusca* is also dichromatic. Hence, the trait is far more common among monogamous primates than among primates generally. For a brief review of possible explanations, see S. Blaffer Hrdy and J. Hartung, "The evolution of sexual dichromatism among primates," *American Journal of Physical Anthropology* 509(3):450 (abstract) (1979).

20. R. Tenaza, "Territory and monogamy among Kloss' gibbons (*Hylobates klossii*) in Siberut Island, Indonesia," *Folia Primatologica* 24:60–80 (1975).

21. W. A. Mason, "Social organization of the South American monkey *Callicebus moloch:* a preliminary report," *Tulane Studies in Zoology* 13:23–29 (1966).

22. Jonathan Pollock's complete study is described in his 1975 Ph.D. thesis, "Social behavior and ecology of *Indri indri,*" presented to the University of London; additional references can be found in his 1977 article, "The ecology and sociology of feeding in *Indri indri,*" in *Primate Ecology,* ed. T. Clutton-Brock (New York: Academic Press), and in "Female dominance in *Indri indri,*" *Folia Primatologica* 31:143–164 (1979).

23. I first heard the argument that monogamy arose in response to adversity in the course of anthropological discussions about the prevalence of monogamy among hunter–gatherer peoples. According to this argument, a woman needed a man to bring food home, but he could rarely support more than one wife. A formal variation of this argument was first presented by E. O. Wilson, who listed three sets of conditions which would account for monogamy, among them a "physical environment so difficult that two adults are needed to cope with it." The other two conditions were a territory containing scarce resources and requiring two animals to defend it, and a situation in which early breeding is advantageous and where previously mated pairs have a head start. *Sociobiology: The New Synthesis* (Cambridge: Harvard University Press, 1975), p. 330. See also Wittenberger and Tilson (note 3) for an extension and reformulation of Wilson's three explanations for monogamy.

24. I. Tattersall and W. R. Sussman, "Observations on the ecology and behavior of the mongoose lemur *Lemur mongoz mongoz* Linnaeus (Primates, Lemuriformes) at Ampijoroa, Madagascar," *Anthropological Papers of the American Museum of Natural History* 52(4):195–216 (1975).

25. P. Wright, "Home range, activity pattern, and agonistic encounters of a group of night monkeys (*Aotus trivirgatus*) in Peru," *Folia Primatologica* 29:43–55 (1978).

26. Of the facultatively monogamous primates, none is quite so variable in its habitats as the mongoose lemur of Madagascar and,

more recently, of the Comoro Islands, where it was introduced by humans. In most respects, these small brown prosimians resemble other members of their genus who live in large, complex social groups. *Lemur mongoz* even sounds the same when encountered in the forest: it utters a noise rather like a creaking door. But unlike other lemurs, which are invariably diurnal and polygynous, mongoose lemurs are sometimes nocturnal and are often found in small family groups. Except for preliminary studies of mongoose lemurs, we know almost nothing about them or their variability from place to place and season to season. We do know, however, that they are not territorial and they are apparently the only monogamous primate apart from *Aotus* which is not.

4. A Climate for Dominant Females

1. Alison Jolly, *Lemur Behavior: A Madagascar Field Study* (Chicago: University of Chicago Press, 1966), pp. 69 ff.

2. Alison Jolly, *The Evolution of Primate Behavior* (New York: Macmillan, 1972), p. 185.

3. Alison Richard, "Patterns of mating in *Propithecus verreauxi verreauxi*," in *Prosimian Biology*, eds. R. D. Martin, G. A. Doyle, and A. C. Walker (London: Duckworth, 1974).

4. Frank V. DuMond, "The squirrel monkey in a semi-natural environment," in *The Squirrel Monkey*, eds. L. Rosenblum and R. W. Cooper (New York: Academic Press, 1968). See also John D. Baldwin, "The social organization of a semifree-ranging troop of squirrel monkeys *(Saimiri sciureus)*," *Folia Primatologica* 14:23–50 (1971). The work of Baldwin and DuMond provides the most extensive published accounts of squirrel monkey behavior, and there is general agreement from both studies that adult females constitute the core of the social group, and to some extent regulate membership in it. However, caution is called for since both studies were carried out in the "seminatural" enclosure at Monkey Jungle. Robert C. Bailey (personal communication) pointed out that in the course of his observations of wild squirrel monkeys in Brazil, females were not always dominant to males at food sources. Pending further information, I accept the published studies at face value, but wish to alert readers to the fact that the full story about squirrel monkeys, once it is known, may be more complicated than any of us yet realize.

5. Thelma E. Rowell, "Reproductive cycles of the talapoin monkey *(Miopithecus talapoin)*," *Folia Primatologica* 28:188–202 (1977), and references therein. A. Gautier-Hion, "Étude du cycle annuel de réproduction du talapoin *(Miopithecus talapoin)* vivant dans son milieu naturel," *Biologica Gabonica* 4:163–173 (1968), and "L'organisation sociale d'une bande de talapoins *(Miopithecus*

talapoin) dans le nord-est du Gabon," *Folia Primatologica* 12:116–141 (1970), and personal communication from Dr. Gautier-Hion.

6. Jaclyn H. Wolfheim, "Sex differences in behavior in a group of captive talapoin monkeys (*Miopithecus talapoin*)," *Behaviour* 63(1–2):110–128 (1977).

7. Jane Lancaster, "Play-mothering: the relation between juvenile females and young infants among free-ranging vervet monkeys (*Cercopithecus aethiops*)," *Folia Primatologica* 15:161–182 (1971); and S. Blaffer Hrdy, "Care and exploitation of nonhuman primate infants by conspecifics other than the mother," in *Advances in the Study of Behavior*, vol. 6 (New York: Academic Press, 1976).

8. Elwyn Simons, and others, have referred to *Miopithecus talapoin* as the smallest and most primitive living African monkey, but in a reconsideration of the status of the talapoin, Eric Delson writes that "far from being primitive, the talapoin is a dentally typical member of the derived Cercopithecini, which have lost M_3 hypoconulids [enamel structures on the third molar], reduced lateral flare, and generally elongated cheek-teeth." E. Delson, "Fossil colobine monkeys of the circum-Mediterranean region and the evolutionary history of the Cercopithecidae (Primates, Mammalia)," Ph.D. thesis presented to Columbia University (1973), p. 639.

5. The Pros and Cons of Males

1. Direct care of infants, that is feeding or carrying them, is reported for 9–10 percent of mammalian genera, contrasted with 40 percent of primate genera which exhibit direct care. Indirect care, such as general protection of a group which contains infants, is virtually universal in social primates. For a review of male care in mammals generally, see Devra G. Kleiman and James R. Malcolm, "The evolution of male parental investment in mammals," in *Parental Care in Mammals*, eds. D. J. Gubernick and P. H. Klopfer (New York: Plenum Press, in press). For reviews of male care among primates, see Sarah Blaffer Hrdy, "The care and exploitation of nonhuman primate infants by conspecifics other than the mother," *Advances in the Study of Behavior* 6:101–158 (1976); William Redican, "Adult male–infant interactions in nonhuman primates;" and M. W. West and M. Konner, "The role of the father: an anthropological perspective," both in *The Role of the Father in Child Development*, ed. M. D. Lamb (New York: John Wiley, 1976).

2. C. O. Lovejoy, "The origin of Man," *Science* 211:341–350 (1981).

3. Obviously, there are exceptions to the rule that the presence of males detracts from food available for mothers and offspring. For example, some monkeys feed on swarms of insects flushed into

flight and it could be argued that males compensate for what they eat by flushing additional food. Nevertheless, for most primates, favored foods are ripe fruits, young leaves, gum exuded from cuts in trees, and so on—all finite commodities.

4. Troops of Hanuman langurs occasionally engage in intertroop encounters during periods when neither troop contains a male. Sarah Blaffer Hrdy, *The Langurs of Abu: Female and Male Strategies of Reproduction* (Cambridge: Harvard University Press, 1977). Female prominence in territorial defense does not, however, extend to all species. Among gorillas, for example, females apparently depend on the silverback male to protect feeding areas from exploitation by other groups. David Watts, "Feeding ecology of mountain gorillas," paper presented at the Forty-ninth Annual Meeting of Physical Anthropologists, Niagara Falls, April 16–19, 1980.

5. The baboons at issue belonged to Alto's troop, a group which has been under observation by Stuart and Jeanne Altmann and Glenn Hausfater since 1971. It is because of their long hours of painstaking habituation that a stranger such as myself could walk amidst well-known individuals. I am grateful to the Altmanns, and particularly to Dr. Hausfater, my host, for the opportunity to spend time with this troop.

6. For one of only a few extant eyewitness accounts of a pair of baboons killing a leopard, see Eugene Marais' somewhat eccentric book *My Friends the Baboons* (London: Blond and Briggs, 1939; rpt. 1975), p. 44.

7. See especially T. W. Ransom and B. S. Ransom, "Adult male–infant relations among baboons (*Papio anubis*)," *Folia Primatologica* 16:179–195 (1971); and Jeanne Altmann, *Baboon Mothers and Infants* (Cambridge: Harvard University Press, 1980).

8. This account is taken from H. Abegglen and J.-J. Abegglen, "Field observations of a birth of hamadryas baboons," *Folia Primatologica* 25:53–56 (1976). A similar account of a savanna baboon male (*Papio ursinus*) attending a delivery can be found in Marais, *My Friends the Baboons* (note 6), p. 50.

9. This dramatization of events in the lives of langurs at Mount Abu is based on a real-life situation described in my book *The Langurs of Abu* (note 4). I could not bear to write the same material in the same way still one more time. Obviously, though, there are drawbacks as well as advantages to such a novelistic treatment. The most serious objection to my effort to see the world from the perspective of the langurs themselves, presenting them as seemingly conscious individuals caught up in a complex breeding system, is that I go too far, leaving the impression that langurs are cost-accountants analyzing their options. Any reader serious enough to be reading this footnote is cautioned, then, that the "mentality" at

work in this drama does not belong to any one individual. The real protagonist is natural selection itself—the disproportionate representation of genes. Over time, the prevailing genes will be those that predispose animals to respond to specific situations in particularly advantageous ways.

10. Between 1971 and 1979 the population of monkeys around Mount Abu town grew by 12 percent.

11. The hypothesis that infant killing by adult males is an evolved reproductive strategy is discussed in Sarah Blaffer Hrdy, "Infanticide as a primate reproductive strategy," *American Scientist* 65:40–49 (1977). A recent computer simulation of the langur situation by Michael Chapman and Glenn Hausfater, entitled "The reproductive consequences of infanticide in langurs: a mathematical model," *Behavioral Ecology and Sociobiology* 5:227–240 (1979), spells out just what the advantages for infanticidal males would be in populations with different durations in average male tenure.

12. Not all primatologists agree that infanticide is a male reproductive strategy which evolved through sexual selection. Critics of this hypothesis claim that infanticide is abnormal behavior and must be attributed to social pathology. The clearest statements of this position derive from the published writings of Phyllis Dolhinow, Richard Curtin, and Jane Boggess. Dolhinow, for example, writes that "in normal troops langur males do not kill infants." Deaths due to infanticide are abnormal and represent "destruction—not adaptation." See letters to the editors, *American Scientist* 65:266 (1977), and subsequent publications by Curtin and Dolhinow entitled, "Primate social behavior in a changing world," *American Scientist* 66(4):468–475 (1978), and "Infanticide among langurs—a solution to overcrowding?" *Science Today* 13(7):35–41 (1979), and by Curtin in "Langur social behavior and infant mortality," *Kroeber Anthropological Society Papers* 50:27–36 (1977). More detailed criticisms of the sexual selection hypothesis can be found in Boggess, "Troop male membership changes and infant killing in langurs (*Presbytis entellus*)," *Folia Primatologica* 32:65–107 (1979), and in Christian Vogel, "Der Hanuman-langur (*Presbytis entellus*), ein Parade-Exempel für die theoretischen Konzepte der 'Soziobiologie'?" *Verhandlungen der Deutschen Zoologischen Gesellschaft* 197:73–89 (1979).

13. For a discussion of the controversy surrounding infanticide and a review of the relevant literature, see Sarah Blaffer Hrdy, "Infanticide among animals: a review, classification and examination of the implications for the reproductive strategies of individuals," *Journal of Ethology and Sociobiology* 1(1):13–40 (1979).

14. S. M. Mohnot, "Some aspects of social change and infant-killing in the Hanuman langur (*Presbytis entellus*) (Primates: Cercopithecidae) in western India," *Mammalia* 35:175–198.

15. Yukimaru Sugiyama, "On the social change of Hanuman langurs (*Presbytis entellus*) in their natural conditions," *Primates* 6:381–418 (1965).

16. Rasanayagam Rudran, "Adult male replacement in one-male troops of purple-faced langurs (*Presbytis senex senex*) and its effect on population structure," *Folia Primatologica* 19:166–192 (1973).

17. K. Wolf, "Social change and male reproductive strategy in silvered leaf-monkeys, *Presbytis cristata*, in Kuala Selangor, Peninsular Malaysia," a paper presented at the Forty-ninth Annual Meeting of Physical Anthropologists, Niagara Falls, April 17–19, 1980. An earlier report by Kathy Wolf and John Fleagle documented male takeovers followed by disappearance of infants, but not witnessed killings; "Adult male replacement in a group of silvered leaf monkeys (*Presbytis cristata*) at Kuala Selangor, Malaysia," *Primates* 18(4):949–955 (1977).

18. John Oates, "The social life of a black-and-white colobus monkey, *Colobus guereza*," *Zeitschrift für Tierpsychologie* 45:1–60 (1977).

19. Thomas Struhsaker, "Infanticide and social organization in the redtail monkey (*Cercopithecus ascanius schmidti*) in the Kibale Forest, Uganda,"*Zeitschrift für Tierpsychologie* 45:75–84 (1977).

20. A. Galat Luong and Gérard Galat, "Consequences comportementales de perturbations sociales repetées sur une troupe de Mones de Lowe *Cercopithecus campbelli lowei* de Côte d'Ivoire," *Terre et Vie* 33:4–57 (1979).

21. T. M. Butynski, "Infanticide in the blue monkey, *Cercopithecus mitis stuhlmanni*, in the Kibale Forest, Uganda," preliminary program of the Eighth International Congress of Primatology, Florence, Italy, July 1980, and personal communication from T. Butynski. Curt Busse and W. J. Hamilton, III, "Why do adult male chacma baboons carry infants when confronting higher ranking males?" paper presented at the Eighth International Congress of Primatology, Florence, Italy, July 7–17, 1980.

22. Rasanayagam Rudran, "Infanticide in red howlers (*Alouatta seniculus*) of northern Venezuela," paper presented at the Seventh International Congress of Primatologists, Bangalore, India, January 8–12, 1979; "The demography and social mobility of a red howler (*Alouatta seniculus*) population in Venezuela," in *Vertebrate Ecology in the Northern Neotropics*, ed. John Eisenberg (Washington: Smithsonian, 1979). Ranka Sekulic, "Infanticide without male takeover in red howlers (*Alouatta seniculus*)," (in prep.).

23. David Bygott, "Cannibalism among wild chimpanzees," *Nature* 238:410–411 (1974).

24. Jane Goodall, "Infant killing and cannibalism in free-living

chimpanzees," *Folia Primatologica* 28:259–282 (1977). See also A. Suzuki, "Carnivority and cannibalism among forest-living chimpanzees," *Journal of the Anthropological Society of Nippon* 79:30–48 (1971), and T. Nishida, "Predatory behavior among wild chimpanzees of the Mahale mountains," *Primates* 20(1):1–20 (1979).

25. Dian Fossey, "The behavior of the mountain gorilla," Ph.D. thesis presented to Cambridge University (1976).

6. Competition and Bonding among Females

1. Allomaternal care of infants, or "aunting," as it used to be called, was first described by Thelma Rowell, Robert Hinde, and Yvette Spencer-Booth for a rhesus monkey colony kept at Madingley, near Cambridge, England. See " 'Aunt'–infant interactions in captive rhesus monkeys," *Journal of Animal Behaviour* 12:219–226 (1964). A decade later, I reviewed the literature on alloparental care of infants among primates and attempted to analyze infant sharing from the perspective of each of the parties involved: the mother, the infant, and the allomother. See S. Blaffer Hrdy, "Care and exploitation of nonhuman primate infants by conspecifics other than the mother," *Advances in the Study of Behavior* 6:101–158 (1976). Status differentials in the willingness of different macaque mothers to give up their infants have been examined by Joan Silk, "Kidnapping and female competition among captive bonnet macaques," *Primates* 21(1):100–110 (1980).

2. The hypothesis that infant sharing gives the mother freedom to forage is supported by recent field research on vervet monkeys. Food intake for recent mothers carrying their infants is lower than for mothers whose infants are being carried by allomothers. Patricia Whitten, "Reproductive strategies of females among wild vervet monkeys," Ph.D. thesis, Harvard University (in prep.).

3. Adoptions have only rarely been reported among wild primates, and the actual incidence of adoption is not known. Several species of primates (rhesus macaques, spider monkeys, humans) have exhibited a capacity to lactate spontaneously after exposure to foster infants. Obviously, such capacities increase the likelihood of survival of orphaned infants, but it is just not known what role this phenomenon plays in nature. See Alejandro Estrada and James Patterson, "A case of adoption in a captive group of Mexican spider monkeys (*Ateles geoffroyi*)," *Primates* 21(1):128–129 (1980). A case of adoption among wild Hanuman langurs in the Himalayas was observed by Jane Boggess (forthcoming). Generally speaking, however, allomothers avoid suckling the infants they borrow from other females.

4. The first convincing data from the field to support the hypothesis that allomothers were indeed primarily inexperienced fe-

males practicing for motherhood were obtained by Jane Lancaster, and are described in "Play-mothering: the relations between juvenile females and young infants among free-ranging vervet monkeys (*Cercopithecus aethiops*)," *Folia Primatologica* 15:161–181 (1971). To date the most detailed studies of infant sharing are for langur monkeys, and support Lancaster's conclusion. See S. Blaffer Hrdy, "The puzzle of langur infant-sharing" in *The Langurs of Abu* (Cambridge: Harvard University Press, 1977). The question of why infant sharing should be especially common among colobine species is addressed in James McKenna's "The evolution of allomothering behavior among colobine monkeys: function and opportunism in evolution," *American Anthropologist* 81 (4):818–840 (1979).

5. Duane Quiatt, "Aunts and mothers: adaptive implications of allomaternal behavior among nonhuman primates," *American Anthropologist* 81 (2):310–319 (1979). Intertroop kidnappings among Indian langurs have been reported by Y. Sugiyama at Dharwar, S. M. Mohnot at Jodhpur, and Daniel and Sarah Hrdy at Mount Abu, reviewed in *Langurs of Abu* (note 4), pp. 153, 225–227.

6. Evidence that the presence of a dominant female altered hormone levels in subordinates among common marmosets and the saddle-back tamarin was reviewed in Chapter 3 (note 11). Similar studies with talapoin monkeys are described in L. A. Bowman, S. R. Dilley, and E. B. Keverne, "Suppression of oestrogen-induced LH surges by social subordination in talapoin monkeys," *Nature* 275:56–58 (1978). Recent literature on reproductive inhibition is reviewed by Samuel K. Wasser and David Barash, "Reproductive inhibition among female mammals" (in press).

7. Anne B. Clark, "Sex ratio and local resource competition among a prosimian primate," *Science* 201:163–165 (1978).

8. In an elegant experiment involving wild hamadryas baboons, Hans Kummer, W. Gotz, and Walter Angst documented the existence of male inhibitions toward females in harems of other males belonging to the same herd; see "Triadic differentiation: an inhibitory process protecting pair bonds in baboons," *Behaviour* 49:62–87 (1974). Similar inhibitions apparently prevent gelada males from attempting to interact with females in other units. Robin Dunbar and Patsy Dunbar, *Social Dynamics of Gelada Baboons*, Contributions to Primatology, vol. 6 (Basel: S. Karger, 1975), p. 44.

9. Hans Kummer, *Social Organization of Hamadryas Baboons* (Chicago: University of Chicago Press, 1968). Interestingly, the only other primates with a comparable social structure are gorillas, which live in a completely different, forest, habitat. Like hamadryas baboons, female gorillas "merely tolerate" other females, and affiliative behaviors among females are uncommon. The focus of social attention is the male, who has gathered his harem of unrelated females together from diverse sources. See especially Alexander Harcourt,

"Social relationships among adult female mountain gorillas," *Animal Behaviour* 27:251–264 (1979).

10. Personal communication from Dr. Joseph Popp.

11. Ueli Nagel, "A comparison of anubis baboons, hamadryas baboons and their hybrids at a species border in Ethiopia," *Folia Primatologica* 19:104–165 (1973).

12. Robin and Patsy Dunbar (note 8); *Ecological and Sociological Studies of Gelada Baboons,* ed. Masao Kawai, Contributions in Primatology, vol. 16 (Basel: S. Karger, 1979). Preliminary studies of gelada baboons were undertaken in the sixties by John Crook.

13. See especially U. Mori, "Individual relationships within a unit," in *Ecological and Sociological Studies of Gelada Baboons* (note 12).

14. R. Dunbar and P. Dunbar, "Dominance and reproductive success among female gelada baboons," *Nature* 206:251–252 (1979). Robin Dunbar reanalyzes these data and expands on the topic in his later paper, "Determinants and evolutionary consequences of dominance among female gelada baboons," *Behavioral Ecology and Sociobiology* 7:253–265 (1980).

15. Lorraine Roth Herrenkohl, "Prenatal stress reduces fertility and fecundity in female offspring," *Science* 206:1097–1099 (1979), and references therein.

16. This pattern has been reported by several observers, but to my knowledge the phenomenon was first described by Thelma Rowell, "Female reproductive cycles and social behaviour in primates," *Advances in the Study of Behavior* 4:69–105 (1972). For the effects of fighting upon menstrual cycles, see Rowell's earlier article, "Baboon menstrual cycles affected by social environment," *Journal of Reproduction and Fertility* 21:133–141 (1970), and Irven DeVore, "The social behavior and organization of baboon troops," Ph.D. thesis presented to the University of Chicago (1962).

17. E. B. Keverne, "Social organization and its consequences for neuroendocrine status in heterosexual groups of talapoin monkeys," *Antropologica Contemporanea: Abstracts of the Seventh International Congress of Primatology* 3(2):221 (1980).

18. An excellent review of the literature on the relationship between nutritional factors and fertility can be found in S. Gaulin and M. Konner, "On the natural diet of primates, including humans," in *Nutrition and the Brain,* vol. 1, eds. R. J. Wurtman and J. J. Wurtman (New York: Raven Press, 1977). Extensive documentation for deferment of reproduction among rodents pending more favorable environmental conditions can be found in F. H. Bronson, "The reproductive ecology of the house mouse," *Quarterly Review of Biology* 54(3):265–299 (1979).

19. Jane Goodall, "Infant-killing and cannibalism in free-living chimpanzees," *Folia Primatologica* 28:259–282 (1977).

20. Yukimaru Sugiyama, "Life history of male Japanese monkeys," *Advances in the Study of Behavior* 7:255–284 (1976).

21. The original report of the macaque dominance system was published in Japanese by S. Kawamura, "The matriarchal social order in the Minoo-B troop: a study on the rank system of Japanese macaques," *Primates* 1:149–156 (1958). Kawamura's findings were later substantiated in research by Naoki Koyama, "On dominance rank and kinship of a wild Japanese monkey in Arashiyama," *Primates* 8:189–216 (1967). Subsequent work by Donald Sade and others has confirmed the existence of a similar organization among rhesus macaques. See especially Sade's "Determinants of dominance in a group of free-ranging rhesus monkeys," in *Social Communication among Primates,* ed. S. Altmann (Chicago: University of Chicago Press, 1967).

22. S. R. Schulman and B. Chapais, "Reproductive value and rank relations among macaque sisters," *American Naturalist* 115(4):580–593 (1980).

23. See especially David Taub, "Age at first pregnancy and reproductive outcome among colony-born squirrel monkeys (*Saimiri sciureus* Brazilian),"*Folia Primatologica* 33:262–272 (1980), and Ken Glander, "Reproduction and population growth in free-ranging mantled howler monkeys," *American Journal of Physical Anthropology* 53:25–36 (1980). Similarly high mortalities for first- and second-born infants are also reported for colony-dwelling rhesus macaques (note 24). The outstanding question for this hypothesis is whether or not rising in rank at about the time of first pregnancy actually enhances the chances that a young mother's offspring will survive. At present, sample sizes which would permit us to compare pregnancy outcome for primiparae who rise in rank with those who for some reason fail to are too small to allow any conclusions to be made. These ideas are discussed further in a manuscript by myself and Barbara Smuts entitled "Ascent of the younger daughter: value or vulnerability" (in prep.).

24. Lee Drickamer, "A ten-year summary of reproductive data for free-ranging *Macaca mulatta,*" *Folia Primatologica* 21:61–80 (1974). D. Sade, K. Cushing, P. Cushing, J. Dunaif, A. Figueroa, J. Kaplan, C. Lauer, D. Rhodes, and J. Schneider report similar trends in "Population dynamics in relation to social structure on Cayo Santiago," *Yearbook of Physical Anthropology* 20:253–262 (1976).

25. Glenn Hausfater, "Long-term consistency of dominance relations in baboons (*Papio cynocephalus*)," paper presented at the Eighth International Congress of Primatology, Florence, Italy, July 1980. See also G. Hausfater and D. F. Watson, "Social and reproductive correlates of parasite and ova emissions by baboons," *Nature* 262:688–689 (1976).

26. C. B. Koford, "Ranks of mothers and sons in bands of rhesus monkeys," *Science* 141:356–357 (1963). Hausfater (1980) reports a similar situation for baboons (note 25).

27. Joan B. Silk, Amy Samuels, and Peter Rodman, "Rank, reproductive success and skewed sex ratios in *Macaca radiata*," paper presented at the Forty-ninth Meeting of the American Association of Physical Anthropologists, Niagara Falls, April 16–19, 1980. For similar data for wild *Papio cynocephalus*, see Jeanne Altmann, "The ecology of motherhood," Ph.D. thesis presented to the University of Chicago (1979).

28. Wolfgang Dittus, "The social regulation of population density and age–sex distribution in the toque monkey," *Behaviour* 63:281–321 (1977). Akio Mori, "An experiment on the relation between the feeding speed and the caloric intake through leaf eating in Japanese monkeys," *Primates* 20(2):185–195 (1979).

29. Clara B. Jones, "The functions of status in the mantled howler monkey, *Alouatta palliata* Gray: intraspecific competition for group membership in a folivorous neotropical primate," *Primates* 21(3):389–405. The recent finding by Ken Glander that rank and reproductive success among howler monkeys are not well correlated lends support to Jones's model. Glander found that middle-ranking howler females had a better rate of infant survivorship than did the very young (often primiparous) females at the top of the hierarchy. Although sample sizes in this study are still small and hence the findings preliminary, it appears that long-term group membership rather than rank position per se may be the critical variable for female reproductive success in this species; see "Reproduction and population growth in free-ranging howler monkeys" (note 23).

30. Sarah Blaffer Hrdy and Daniel B. Hrdy, "Hierarchical relations among female Hanuman langurs (Primates: Colobinae, *Presbytis entellus*)," *Science* 193:913–915 (1976).

31. Leanne Nash, "Troop fission in free-ranging baboons in the Gombe Stream National Park, Tanzania," *American Journal of Physical Anthropology* 44:63–77 (1976). Diane Chepko-Sade and D. S. Sade, "Patterns of group splitting within matrilineal kinship groups," *Behavioral Ecology and Sociobiology* 5:67–86 (1979), and the related article by Diane Chepko-Sade and T. J. Olivier, "Coefficient of genetic relationship and the probability of intragenealogical fission in *Macaca mulatta*," *Behavioral Ecology and Sociobiology* 5:263–278 (1979). Koyama describes comparable troop fission among *Macaca fuscata* (note 33).

32. See review of currrent literature by Dennis Chikazawa, T. P. Gordon, C. A. Bean, and I. S. Bernstein, "Mother-daughter dominance reversal in rhesus monkeys (*Macaca mulatta*)," *Primates* 20(2):301–305 (1979).

33. H. Marsden, "Agonistic behaviour of young rhesus monkeys

after changes induced in social rank of their mothers," *Animal Behaviour* 16:38–44 (1968); and especially Naoki Koyama, "Changes in dominance rank and division of a wild Japanese monkey troop in Arashiyama," *Primates* 11:335–390 (1970), I. Bernstein, "Spontaneous reorganization of a pigtail monkey group," *Proceedings of the Second International Congress of Primatology, Atlanta, Georgia* (Basel: S. Karger, 1969), and H. Gouzoules, "A description of genealogical rank changes in a troop of Japanese monkeys (*Macaca fuscata*)," *Primates* 21(2):262–267 (1980).

34. M. R. A. Chance, G. R. Emory, and R. G. Payne, "Status references in long-tailed macaques (*Macaca fascicularis*): precursors and effects of a female rebellion," *Primates* 18(3):611–632 (1977).

35. Leanne T. Nash, "Parturition in a feral baboon (*Papio anubis*)," *Primates* 15(2–3):279–285 (1974).

36. R. Seyfarth, "Social relationships among adult male and female baboons, pt. 2: behaviour throughout the menstrual cycle," *Behaviour* 64(3–4):227–247 (1978).

37. Dorothy Cheney, "The acquisition of rank and development of reciprocal alliances among free-ranging immature baboons," *Behavioral Ecology and Sociobiology* 2:303–318 (1977). The work of Cheney and Seyfarth (note 36) differs from many earlier primate studies in the attention they pay to "relationships" and to the history of interactions between intelligent animals living in complex and long-lasting social systems. They follow in their analysis the pioneering approach taken by Robert Hinde and his associates.

38. Personal communication from Barbara Smuts and Nancy Nicolson. A preliminary description of the rebellion at Gilgil, "Effects on social behavior of loss of high rank in wild adult female baboons (*Papio anubis*)," was presented by Smuts at the Annual Meeting of the Animal Behavior Society, Fort Collins, Colorado, June 9–13, 1980.

39. Earlier efforts included J. H. Crook and J. S. Gartlan, "Evolution of primate societies,"*Nature* 210:1200–1203 (1966), and Woodrow W. Denham, "Energy relations and some basic properties of primate social organization," *American Anthropologist* 73:77–95 (1971). But Wrangham's attention to evolutionary theory as well as his familiarity with ecological studies by Jarman, Bell, and others make his analysis the most exciting to date. See Richard Wrangham, "The behavioral ecology of chimpanzees at Gombe Stream National Park, Tanzania," Ph.D. thesis presented to Cambridge University (1975), and especially "On the evolution of ape social systems," *Biology and Social Life: Social Sciences Information* 18(3):335–368 (1979). Ideas presented in these papers are carried further and made more general in "An ecological model of female-bonded primate groups," *Behaviour* 75(3–4):262–300.

40. Christian Vogel, "Hanuman as an object for anthropologists—field studies of social behavior among the gray langurs of India," in *German Scholars on India: Contributions to Indian Studies,* vol. 1 (Varanasi, India: Chowkhamba/Sanskrit Series Office, 1973), p. 363. This is by no means an isolated example; see also Carol Cronin, "Dominance relations and females," in *Dominance Relations: An Ethological View of Human Conflict and Social Interaction,* and the references therein (New York: Garland, 1980), p. 302.

41. Virginia Abernethy, "Female hierarchy: an evolutionary perspective," in *Female Hierarchies,* eds. Lionel Tiger and Heather Fowler (Chicago: Beresford Book Service, 1978), pp. 127–129. Abernethy's views are echoed in the recent paper by Cronin (note 40): "The important point is that it is male competition that must be brought under control to keep the social group functioning in an orderly way. Besides competing for food, resting places, and grooming or other forms of attention, males also compete for sexual access to females (females do not compete for similar exclusive rights to males). While virtually every female is assured of leaving offspring, there is probably a great range in the number of offspring a given male might leave . . . Female groups, especially in rhesus and baboon societies, tend to be much noisier and more disorganized than male groups . . . In a variety of nonhuman primate species, then, competition among males is more crucial (since leaving offspring is at stake) . . . The question then arises whether or not male primates (including humans) have some biological preadaptation for competition." Relying largely on conclusions about nonhuman primates formulated in the 1960s before much reliable information on female primates was available, both authors fall into the trap of imagining that only males compete. I do not mean to single out particular authors for criticism. Rather, I have selected these examples because they set forth with unusual clarity positions which represent a prevailing view among social scientists.

42. J. Shepher and L. Tiger, "Female hierarchies in a kibbutz community," in *Female Hierarchies* (note 41), p. 246.

43. I have benefited greatly in this brief review of studies about women done by social psychologists and anthropologists from several discussions with John and Beatrice Whiting. The Whitings share my interest in competition among women and are currently designing methods to measure competitiveness among adult women which would permit cross-cultural comparison of the phenomenon. Among the ethnographers who deal in passing with the question of competition among women, see especially Hortense Powdermaker, *Copper Town* (New York: Harper and Row, 1962), pp. 207–214. See also note 53 in Chapter 8 for additional ethnographic examples. A recent study of girls at summer camp illustrates some of the difficulties in

trying to quantify competitive interactions; R. C. Savin-Williams, "Social interactions of adolescent females in natural groups," in *Friendship and Social Relations among Children*, eds. H. Goot, T. Chapman, and J. Smith (Sussex, England: Wiley, in press). By and large, however, the emphasis in recent years has been on documenting the degree to which women cooperate. See, for example, *Women United, Women Divided: Comparative Studies of Ten Contemporary Societies*, eds. Patricia Caplan and Janet Bujra (Bloomington: Indiana University Press, 1979).

44. William L. O'Neill, "Women in politics," in *Female Hierarchies* (note 41), p. 219. O'Neill is also the author of *Everyone Was Brave: The Rise and Fall of Feminism in America* (New York: Quadrangle Books, 1969), which provides further documentation for the statements he makes in his essay on women in politics.

7. The Primate Origins of Female Sexuality

1. N. McWhirter and R. McWhirter, *Guinness Book of World Records* (New York: Sterling, 1975). Napoleon Chagnon, "Is reproductive success equal in egalitarian societies?" in *Evolutionary Biology and Human Social Behavior*, eds. N. Chagnon and W. Irons (North Scituate, Mass.: Duxbury Press, 1979), p. 379. *Boston Globe*, February, 7, 1975.

2. In a review of what one author considered to be the relevant primate data, one recent textbook reads that "only males are directly involved in differential selection among rhesus and probably all the terrestrial or semiterrestrial primates." Daniel G. Freedman, *Human Sociobiology: A Holistic Approach* (New York: The Free Press, 1979), p. 33.

3. M. Daly and M. Wilson, *Sex, Evolution and Behavior* (North Scituate, Mass.: Duxbury Press, 1978), pp. 58–59. Of the recent textbooks on sociobiology, this balanced account is probably the most successful. It is a good example for my purpose because it cannot be dismissed as "extreme." See also the short article by the same authors entitled "Sex and strategy," *New Scientist*, January 4, 1979, pp. 15–17.

4. Donald Symons, *The Evolution of Human Sexuality* (Oxford: Oxford University Press, 1979), p. 261. See popularizations of these same ideas in Scott Morris, "Darwin and the Double Standard," *Playboy Magazine*, August 1979, pp. 109 ff, and, from time to time, in most local newspapers.

5. Long-term data from hunter–gatherers like the !Kung, whose lives are being drastically modified by contact with the outside world, are now out of reach for modern science. Nevertheless, by combining death and fertility rates recorded for a real-life population with a computer simulated facsimile, Howell was able to model what

might have happened in a traditional group over many generations. Nancy Howell, *Demography of the Dobe !Kung* (New York: Academic Press, 1979). Enterprises such as Howell's have a venerable if checkered history. R. A. Fisher, one of the first scientists to apply statistical methods to population genetics and demography, shared her concern with variance in female fertility. For example, Fisher showed that the number of children produced by a woman is significantly correlated with the number of children born to their mothers and grandchildren. *The Genetical Theory of Natural Selection* (1930; rpt. New York: Dover, 1958), pp. 213 ff.

6. Howell's finding that cattlepost women reproduced more held true even if the settled wife was married polygynously. Most bush-living !Kung practice monogamy. Tentatively, Howell (note 5) explains the difference in birth spacing among !Kung and cattlepost women by citing the Frisch hypothesis. According to this hypothesis, a critical ratio of fat to lean (or muscle) is necessary for a woman to reproduce. See Rose Frisch and Janet MacArthur, "Menstrual cycles: fatness as determinant of minimum weight for height necessary for their maintenance or onset," *Science* 185:949–951 (1974). Historical evidence for a direct link between nutrition and female reproductive capacity is reviewed by Rose Frisch in her article, "Population, food intake and fertility," *Science* 199:22–30 (1978). Frisch's hypothesis continues to be hotly contested, but it is interesting to note that the hypothesis receives support from recent studies of other primates in those cases where information on body weight in addition to long-term data on reproduction are available. See for example Akio Mori, "Analysis of population changes by measurement of body weight in the Koshima troop of Japanese monkeys," *Primates* 20(3):371–397 (1979).

7. Jacques Gomila, "Fertility differentials and their significance for human evolution," in *The Role of Natural Selection in Human Evolution,* ed. F. M. Salzano (Amsterdam: North Holland Publishing Co., 1975), and Barry L. Isaac, "Female fertility and marital form among the Mende of Rural Upper Bambara Chiefdom, Sierra Leone," *Ethnology* 19(3):297–313 (1980).

8. The fertility differential between monogamously and polygynously married women is still evident even if special circumstances surrounding Mende marriage patterns are taken into account, namely, that a divorcee or a widow who proved her fertility in her first marriage would be more likely to enter into a monogamous second marriage. Nevertheless, Isaac (note 7) warns that the fertility differential is probably multicausal. Particularly in the case of junior co-wives in polygynous unions, the lower fertility of these socially subordinate women compared to that of monogamously married women may be explained by various social customs. In addition, the introduction of venereal disease in recent times further complicates

interpretation of these data—as it also does in the case of the !Kung (note 5).

9. Wolfgang Dittus, "The social regulation of population density and age–sex distribution in the toque monkey," *Behaviour* 63:281–321 (1977). See also Akio Mori (note 6).

10. Joan Silk, Cathleen Clark-Wheatley, Peter Rodman, and Amy Samuels, "Differential reproductive success among female *Macaca radiata*: a longitudinal analysis" (in press); Dittus (note 9).

11. Among the langurs at Mount Abu, at least four different males usurped control of a particular troop (Hillside troop) in the period between 1971 and 1975. The minimum estimate for infant mortality during this period was 83%. In some cases, there is probably nothing a female can do to save her infant; she is the victim of chance. In other cases, however, maternal efforts to forestall males from killing her infant may be successful. For example, some females leave the troop, taking their infants with them. In one case (described in Chapter 5 of this book) a mother left her infant with the males who had been ousted and returned by herself to the recently usurped troop. Obviously, if individual mothers differ in their ability to keep infants alive, one would expect males to exercise some preference for particular females. Certainly, there exists anecdotal evidence which suggests that dominant males in species such as baboons, macaques, and chimpanzees prefer consortships with older, multiparous and high-ranking females, that is, females generally liable to be more successful than average at rearing offspring. For a review of some of these cases, see Sarah Blaffer Hrdy, *The Langurs of Abu* (Cambridge: Harvard University Press, 1977), pp. 178–183.

12. Desmond Morris, *The Naked Ape* (New York: McGraw-Hill, 1967), pp. 53–55.

13. See the now classic essay by Wolfgang Wickler which foreshadowed a decade of speculation about the naked apes: "Socio-sexual signals and their intra-specific imitation among primates," in *Primate Ethology*, ed. D. Morris (London: Weidenfeld and Nicolson, 1967), pp. 89–189. See also Roger Short, "Sexual selection and its component parts, somatic and genital selection, as illustrated in man and the great apes," *Advances in the Study of Behavior* 9:131–158 (1978).

14. David Barash provides a representative quotation: "Human infants are totally helpless and require the committed attention of one parent, invariably the woman, since she is also adapted to nourish her newborn. It would certainly help if there was a daddy around to hunt, scavenge, defend the female and her child, etc. Given that, during our evolutionary development, offspring were more likely to be successful if they received the committed assistance of at least two adults, selection would favor any mechanism that kept the

adults together. Sex may be such a device, selected to be pleasurable for its own sake, in addition to its procreative functions. This would help explain why the female orgasm seems to be unique to humans; among other animals, reproduction is the only goal, and satisfaction per se is irrelevant. In addition, loss of estrus among humans contributes to sexual consistency that may in turn help maintain a stable pair-bond." From *Sociobiology and Behavior* (New York: Elsevier, 1977), p. 297.

15. See especially Donald Symons, "Copulation as a female service," in *The Evolution of Human Sexuality* (note 4).

16. G. E. Pugh, *The Biological Origin of Human Values* (New York: Basic Books, 1977), p. 248.

17. William H. Masters and Virginia Johnson, *Human Sexual Response* (Boston: Little, Brown, 1966), pp. 158–159.

18. J. D. Weinrich, "Human sociobiology: pair-bonding and resource predictability (effects of social class and race)," *Behavioral Ecology and Sociobiology* 2:91–118 (1977).

19. J. P. Hearne, "The endocrinology of reproduction in the common marmoset *Callithrix jacchus,*" in *The Biology and Conservation of the Callitrichidae,* ed. D. Kleiman (Washington, D.C.: Smithsonian Institution, 1978).

20. Biruté Galdikas, "Orangutan adaptation at Tanjung Puting Reserve: mating and ecology," in *The Great Apes,* eds. D. Hamburg and E. McCown (Menlo Park: Benjamin Cummings, 1979), pp. 195–233.

21. See for example R. Nadler, "Sexual cyclicity in captive lowland gorillas," *Science* 189:813–814 (1975); J. MacKinnon, "Reproductive behavior in wild orangutan populations," in *Great Apes* (note 20), pp. 257–273, and especially E. S. Savage-Rumbaugh and B. J. Wilkerson, "Socio-sexual behavior in *Pan Paniscus* and *Pan troglodytes:* a comparative study," *Journal of Human Evolution* 7:327–344 (1978).

22. J. A. R. A. M. van Hooff, "A comparative approach to the phylogeny of laughter and smiling," in *Non-verbal Communication,* ed. R. A. Hinde (Cambridge: Cambridge University Press, 1972), pp. 209–241; Frans B. M. de Waal and A. van Roosmalen, "Reconciliation and consolation among chimpanzees," *Behavioral Ecology and Sociobiology* 5:55–66 (1979).

23. Carol Tavris and Susan Sadd, *The Redbook Report on Female Sexuality* (New York: Delacorte Press, 1977), p. 70. These data were compiled from a survey of 100,000 married women who responded to a questionnaire in *Redbook Magazine.* Inevitably, biases enter in. Although useful, it is well also to keep in mind limitations of information collected in this way. In particular, only readers of *Redbook* were likely to see the questionnaire, and of those, only women with certain attitudes were likely to respond.

24. In 1968 the sociologists J. Richard Udry and Naomi Morris reported a peak both in sexual activity and the likelihood of orgasm which occurred around the time of ovulation, with a second, lesser peak, just prior to menstruation. Their article, "Distribution of coitus in the menstrual cycle," *Nature* 200:593–596 (1968), drew criticism from various quarters. See especially W. H. James, "The distribution of coitus within the human intermenstruum," *Journal of Biosocial Sciences* 3:159–171 (1977). Nevertheless, Udry and Morris's findings accord well with the recent study by David Adams, Anne Burt, and Alice Ross Gold entitled "Rise in female-initiated sexual activity at ovulation and its suppression by oral contraceptives," *The New England Journal of Medicine* 299:1145–1150 (1978). Because it is thought that birth control pills can alter the patterning of sexual desire, it should be noted that none of the subjects in the Adams, Burt, and Gold study were taking oral contraceptives.

25. *New England Journal of Medicine* 300:625–627 (1979). In addition to that dissenting opinion, a critical editorial also appeared in the same issue of the *NEJM* containing the original study by Adams, Burt, and Gold (note 24). Clearly, "expert" opinions remain divided over whether or not there is a correlation between menstrual cycle stage and willingness to engage in sexual activity, and, if there is a correlation, just what the patterning of those mood changes looks like when plotted against the course of the cycle.

In a recent issue of the journal *Signs* devoted to the subject of "Women—sex and sexuality," authors disagree from one article to the next. In an extensive and up-to-date review of the literature, Richard C. Friedman, Stephen W. Hurt, Michael S. Arnoff, and John Clarkin stress the importance of biological factors in addition to psychological and sociocultural factors in explaining mood shifts and behavioral changes women experience in the course of the menstrual cycle; they tend to accept the findings by Adams, Burt, and Gold. "Behavior and the menstrual cycle," *Signs* 5(4):719–738 (1980). But in the same issue, another author, Susan (Contratto) Weisskopf presents an alternate perspective stressing the absence of any consistent relationship between hormone levels (and hence menstrual phase) and sexual behavior. Weisskopf endorses a viewpoint which separates reproduction from sexuality, a view suggested some years ago by the feminist Shulamith Firestone and others. "Maternal sexuality and asexual motherhood," pp. 766–782, but especially pp. 773–775. Oddly enough—since after all, I am attempting to trace female sexuality back to its evolutionary origins—I agree with Weisskopf that it is useful to conceptually separate reproduction (defined as conceiving young) from sexuality (defined as readiness to engage in sexual activities). But, as will become apparent by the end of Chapter 7, I disagree once the concept of "reproduction" is expanded to include not only the conception, gestation, delivery, and suckling of young but also any behavior by a female which increases the likelihood her offspring will survive and which increases her own long-term reproductive success. By this definition, reproduction is integral to the evolution of female sexuality.

26. Carol Worthman, "Psychoendocrine study of human behavior: some interactions of steroid hormones with affect and behavior in the !Kung San," Ph.D. thesis presented to Harvard University (1978).

27. K. R. L. Hall and Irven DeVore, "Baboon social behavior," in *Primate Behavior*, ed. I. DeVore (New York: Holt, Rinehart and Winston, 1965).

28. Desmond Morris, *The Naked Ape* (note 12), pp. 53–55. Similar arguments have been proposed by Sherwood Washburn and Irven DeVore in "Social behavior of baboons and early man," in *The Social Life of Early Man*, ed. S. L. Washburn (Chicago: Aldine, 1961), and by J. Buettner-Janusch in *Origins of Man* (New York: John Wiley, 1966).

29. Jane Beckman Lancaster, "Sex roles in primate societies," in *Sex Differences*, ed. M. S. Teitelbaum (New York: Doubleday, 1977), p. 51.

30. J. O. Ellefson, "Territorial behavior in the common white-handed gibbon *Hylobates lar*," in *Primates*, ed. P. Jay (New York: Holt, Rinehart and Winston, 1968); J. Raemakers, "Synecology of Malaysian apes," Ph.D. thesis presented to Cambridge University, Cambridge, England (1977); David Chivers, *The Siamang in Malaysia: A Field Study of a Primate in a Tropical Rain Forest*, Contributions to Primatology, 4 (Basel: S. Karger, 1974); "The lesser apes," in *Primate Conservation*, ed. H. S. H. Prince Rainier and G. Bourne (New York: Academic Press, 1977).

31. Richard D. Alexander and Katharine Noonan, "Concealment of ovulation, parental care, and human social evolution," in *Evolutionary Biology and Human Social Behavior* (note 1), pp. 436–453.

32. Nancy Burley, "The evolution of concealed ovulation," *American Naturalist* 114(6):835–858 (1979).

33. The lack of any strict correlation between ovulation and sexual behavior in higher primates has been known since the turn of the century when Walter Heape presented an exhaustive report on estrus in mammals entitled, "The 'sexual season' of mammals and the relation of the 'pro-oestrum' to menstruation," *Quarterly Journal of Microscopic Science* 44:1–70 (1900). However, Heape's observations about the intermediate position of the higher primates between humans and other animals have been frequently glossed over in texts and general reviews, with several outstanding excep-

tions. One of these is the now classic overview of mammalian reproduction by Clellan S. Ford and F. Beach, *Patterns of Sexual Behavior* (New York: Harper and Row, 1951). A second is the excellent review of primate reproduction by Thelma Rowell, "Female reproduction cycles and social behavior in primates," *Advances in the Study of Behavior* 4:69–105 (1972). An updated review of these topics can be found in Robert Hinde, *Biological Bases of Human Social Behaviour* (New York: McGraw-Hill, 1974).

34. Jane Lancaster and Richard B. Lee, "The annual reproductive cycle in monkeys and apes," in *Primate Behavior* (note 27), pp. 486–513. C. H. Conway and D. S. Sade, "The seasonal spermatogenic cycle in free-ranging rhesus monkeys," *Folia Primatologica* 3:1–12 (1965).

35. T. H. Clutton-Brock and P. H. Harvey, "Evolutionary rules and primate societies," in *Growing Points in Ethology*, eds. P. P. G. Bateson and R. A. Hinde (Cambridge: Cambridge University Press, 1976), pp. 195–237.

36. James Loy, "Reproduction in patas monkeys: behavioral and physical phenomena related to gestation and parturition," paper presented at the Forty-third Annual Meeting of the American Association of Physical Anthropologists (1974).

37. Among captive pigtailed macaques (*Macaca nemestrina*) and free-ranging rhesus, copulations are more or less evenly distributed across the menstrual cycle. Both receptivity and the number of intromissions are fairly constant, but the "attractiveness" of females to males may vary. Among captive rhesus, Richard Michael and Doris Zumpe have been able to show that the only measure reliably correlated with a particular stage of a female's reproductive cycle is frequency of male ejaculation, which is highest at his partner's midcycle. C. H. Conaway and C. B. Koford, "Estrous cycles and mating behavior in a free-ranging band of rhesus monkeys," *Journal of Mammalogy* 45:577–588 (1965); James Loy, "Peri-menstrual sexual behavior among rhesus monkeys," *Folia Primatologica* 13:286–297 (1970); G. G. Eaton, "Social and endocrine determinants of sexual behavior in simian and prosimian females," *Symposium IVth International Congress of Primatology*, vol. 2: *Primate Reproductive Behavior* (Basel: S. Karger, 1973), pp. 20–35, and especially R. P. Michael and Doris Zumpe, "Rhythmic changes in the copulatory frequency of rhesus monkeys (*Macaca mulatta*) under laboratory conditions," *Journal of Endocrinology* 41:231–246 (1970).

Captive and wild gelada baboons similarly copulate throughout the cycle, but unlike the macaques, gelada males are no more likely to ejaculate at midcycle than at other times. Female geladas, however, are significantly more likely to solicit males in the ten days or so when the bead-like vesicles on the bare skin of a gelada female's chest and perineal region are maximally swollen and pigmented than they are in the ten-day period surrounding menstruation. R. R. Smith and P. F. Credland, "Menstrual and copulatory cycles in the gelada baboon, *Theropithecus gelada," International Zoo Yearbook,* ed. J. S. Olney (London: New York Zoological Society, 1977). R. I. M. Dunbar and P. Dunbar, "The reproductive cycle of the gelada baboon," *Animal Behaviour* 22:203–210 (1974) and R. I. M. Dunbar, "Sexual behavior and social relationships among gelada baboons," *Animal Behaviour* 26:167–178 (1978).

38. R. D. Nadler, "Sexual behavior of captive orangutans," *Archives of Sexual Behavior* 6:457–475 (1977). W. B. Lemmon and M. L. Allen, "Continual sexual receptivity in the female chimpanzee (*Pan troglodytes*)," *Folia Primatologica* 30:80–88 (1978). E. S. Savage-Rumbaugh and B. J. Wilkerson, "Socio-sexual behavior in *Pan paniscus* and *Pan troglodytes:* a comparative study," *Journal of Human Evolution* 7:327–344 (1978).

39. R. D. Nadler, "Sexual cyclicity in captive lowland gorillas," *Science* 189:813–814 (1975).

40. John MacKinnon, "Reproductive behavior in wild orangutan populations," pp. 257–273, and especially Biruté Galdikas, "Orangutan adaptation," both in *The Great Apes,* eds. David A. Hamburg and Elizabeth R. McGown (Menlo Park: Benjamin-Cummings, 1979), pp. 195–233.

41. C. Tutin, "Sexual behavior and mating patterns in a community of wild chimpanzees (*Pan troglodytes*)," Ph.D. thesis presented to the University of Edinburgh (1975), p. 58.

42. Glenn Hausfater, *Dominance and Reproduction in Baboons* (Basel: S. Karger, 1975).

43. C. Tutin (note 41), pp. 58–59.

44. Alexander Harcourt, "Contrasts between male relationships in wild gorilla groups," *Behavioral Ecology and Sociobiology* 5:39–49 (1979).

45. D. Taub, "Female choice and mating strategies among wild barbary macaques (*Macaca sylvanus* L.)," in *The Macaques: Studies in Ecology, Behavior and Evolution,* ed. D. Lindburg (New York: Van Nostrand-Reinhold Co., 1980), p. 316.

46. Among captive rhesus macaques, for example, two distinct periods of increased sexual interaction have been detected: one period coincident with the preovulatory estradiol peak and the other between the sixth to tenth week of pregnancy. C. Bielert, J. A. Czaja, S. Eisele, G. Scheffer, J. A. Robinson, and R. W. Goy, "Mating in the rhesus monkey (*Macaca mulatta*) after conception and its relationship to oestradiol and progesterone levels throughout pregnancy," *Journal of Reproduction and Fertility* 46:179–187 (1976).

47. C. Tutin (note 41), p. 60. Linda Scott, personal communication (1978). D. Lindburg, "The rhesus monkey in North India: an

ecological and behavioral study," in *Primate Behavior,* ed. L. A. Rosenblum (New York: Academic Press, 1971), p. 92.

48. R. L. Trivers, "The logic of female choice," paper presented at the Annual Meeting of the Animal Behavior Society, Seattle, Washington, June 1978.

49. Robert Seyfarth reports from a troop of South African baboons (*Papio ursinus*) that "three adult females 'alternated' preferences between Rocky [the alpha male] and Pierre [the beta male] according to changes in their reproductive states . . . each female associated primarily with Pierre during lactation and Rocky during sexual cycling." R. M. Seyfarth, "Social relationships among adult male and female baboons, pt 2: Behavior throughout the female reproductive cycle," *Behaviour* 64(3–4):227–247 (1978), esp. p. 239.

50. C. Cox and B. J. LeBoeuf, "Female incitation of male competition: a mechanism in sexual selection," *American Naturalist* 111:317–335 (1977).

51. Clutton-Brock and Harvey (note 35), pp. 210–214.

52. C. Jones, "Aspects of reproductive behavior in the mantled howler monkey, *Alouatta palliata* Gray," Ph.D. thesis presented to Cornell University (1978).

53. W. J. Hamilton and P. C. Arrowood, "Copulatory vocalizations of chacma baboons (*Papio ursinus*), gibbons (*Hylobates hoolock*), and humans," *Science* 200:1405–1409 (1978).

54. C. Tutin (note 41), pp. 52–53.

55. It has been known for several years that chimpanzees and gorillas sometimes transfer between troops (see *The Great Apes,* note 40), but only very recently have fieldworkers realized that wild howler monkey females (Jones, note 53) and red colobus monkey females were also immigrating. See T. Struhsaker and L. Leland, "Socioecology of five sympatric monkey species in the Kibale forest, Uganda," *Advances in the Study of Behavior* 9:159–228 (1979); and C. W. Marsh, "Female transference and mate choice among Tana River red colobus," *Nature* 281:568–569 (1979). Female transfer also occurs among Hamadryas baboons, but adult males "kidnap" immature females from their natal groups and in this respect the behavior of *Papio hamadryas* may differ from what is usually meant by "voluntary" transfers involving adults.

56. A similar idea is proposed by Lee Benshoof and Randy Thornhill in "The evolution of monogamy and concealed ovulation in humans," *Journal of Social and Biological Structures* 2:95–106 (1979). A major difference between their paper and the model discussed here is that Benshoof and Thornhill regard concealed ovulation as a uniquely human attribute, while I do not.

57. F. F. Mallory and R. Brooks, "Infanticide and other reproductive strategies in the collared lemming, *Dicrostoyx groenlandicus,*" *Nature* 273:144–146 (1978).

58. J. B. Labov, "Pregnancy blocking in house mice (*Mus musculus*) and other mammals: sociobiological implications and adaptive strategies for females," Ph.D. thesis presented to the University of Rhode Island, Kingston (1979), and "Factors affecting infanticidal behavior in wild male house mice (*Mus musculus*)," *Behavioral Ecology and Sociobiology* 6:297–303 (1980).

59. Several scientists have suggested that induced ovulation occurs in primates. See especially J. H. Clark and M. X. Zarrow, "Influence of copulation on time of ovulation in women," *American Journal of Obstetrics and Gynecology* 109:1083–1085 (1971). Comparative evidence from the genital anatomy of primates suggests that other primates in addition to women may sometimes be spontaneous ovulators. According to the argument outlined by Zarrow and Clark, possession of penile spines—tiny cornified epithelial structures on the glans of the penis first described for rats—tends to be associated with the occurrence of induced ovulation; primates are the only mammalian order with penile spines in which reflex ovulation has not also been reported. To date, penile spines have been reported for lorisiformes, lemurs, spider monkeys, gibbons, chimpanzees, and orangutans. This morphological evidence, plus the fact that induced ovulation is known to sometimes occur in women, are good reasons to keep in mind the possibility that males who copulate with a supposedly nonovulating female may in fact have some fractional chance of inseminating her. See M. X. Zarrow and J. H. Clark, "Ovulation following vaginal stimulation in a spontaneous ovulator and its implications," *Journal of Endocrinology* 40:117–123. In addition, see J. W. Harms, "Fortpflanzungsbiologie," in *Primatologica: Handbuch der Primatenkunde,* eds. H. Hofer, A. H. Schultz, and D. Starck (Basel: S. Karger, 1956), pp. 561–660, for evidence of penile spines in primates. See S. Sevitt, "Early ovulation," *The Lancet* 2:448–450 (1946) for evidence of induced ovulation in women.

60. Taub (note 45), p. 335.

61. Taub (note 45), p. 337.

62. G. Isaac, "The food-sharing behavior of protohuman hominids," *Scientific American* 238(4):90–106 (1978).

8. A Disputed Legacy

1. Bronislaw Malinowski, *The Sexual Life of Savages* (New York: Harcourt, Brace and World, 1929), p. xxiii.

2. Clifford Geertz, "Sociosexology," *New York Review of Books,* January 24, 1980, p. 4.

3. Margaret Mead, *Male and Female* (New York: William Morrow, 1968), p. 219.

4. A. C. Kinsey, W. B. Pomeroy, C. E. Martin, and P. H. Gebhard, *Sexual Behavior in the Human Female* (Philadelphia: W. B.

Saunders, 1953). Linda Wolfe, "The sexual profile of that cosmopolitan girl," *Cosmopolitan,* September 1980, pp. 254–265.

5. Mary Jane Sherfey, *The Nature and Evolution of Female Sexuality* (New York: Vintage Books, 1973), pp. 29, 80. A shorter version of this book appeared earlier in the *Journal of the American Psychoanalytic Association* 14:28–128 (1966).

6. Changes through time in western attitudes toward female sexuality are examined in Lawrence Stone's monumental book, *The Family, Sex and Marriage in England 1500–1800* (New York: Harper and Row, 1977), esp. p. 489.

7. Sherfey, *Evolution of Female Sexuality* (note 5), p. 112.

8. The Harvard psychologist George Goethals was one who welcomed Sherfey's ideas. I first heard of her work in a lecture by him. In his article, "Factors affecting permissive and nonpermissive rules regarding premarital sex," in *Studies in the Sociology of Sex,* ed. J. M. Henslin (New York: Appleton-Century-Croft, 1971), pp. 9–26, Goethals referred to "the brilliant paper by Mary Jane Sherfey, which . . . abolishes once and for all the whole psychoanalytic interpretation of 'female sexuality.' " Sherfey's writing was similarly credited by Shere Hite in *The Hite Report* (New York: Macmillan, 1976). But Sherfey's ideas were not universally acclaimed. A later issue of the *Journal of the American Psychoanalytic Association,* which had originally published Sherfey's ideas, was devoted to papers refuting them. The major criticisms appeared in a paper by Marcel Heiman, "Discussion of Sherfey's paper on female sexuality," *Journal of the American Psychoanalytic Association,* July 1968. A highly critical book review by Irving Singer appeared in the *New York Review of Books* 19(9):29–31 (November 30, 1972). Similarly, there is more than a hint of sarcasm in Donald Symons' rejection of "Sherfeyian" females. "The sexually insatiable woman is to be found primarily, if not exclusively, in the ideology of feminism, the hopes of boys, and the fears of men." According to Symons, "Sherfey is a sexual radical for whom paradise is endless, orgiastic sexual indulgence." From *The Evolution of Human Sexuality* (Oxford: Oxford University Press, 1979), pp. 92, 94.

9. Symons (note 8), pp. 89, 91, 192.

10. Symons (note 8), p. 92.

11. Stone (note 6), p. 676.

12. Desmond Morris, *The Naked Ape* (New York: McGraw-Hill, 1967), p. 66.

13. Mead (note 3), p. 219.

14. Twenty percent of 300 married women interviewed by Fisher, and some 30 percent of those responding to the Hite survey, said that they could regularly climax from intercourse alone, without additional clitoral stimulation. See review of these studies in Hite (note 8).

15. Larry McFarland, "Comparative anatomy of the clitoris," in *The Clitoris,* eds. Thomas Lowry and Thea Snyder Lowry (St. Louis: Warren H. Green, Inc., 1976), pp. 22–34.

16. "The clitoris is a unique organ in the total of human anatomy. Its express purpose is to serve both as a receptor and transformer of sensual stimuli. Thus, the human female has an organ system which is totally limited in physiological function to initiating or elevating levels of sexual tension. No such organ exists within the anatomic structure of the human male." William Masters and Virginia Johnson, *Human Sexual Response* (Boston: Little, Brown and Co., 1966), p. 45.

17. In Roger Short's detailed article, "Sexual selection and its component parts, somatic and genital selection, as illustrated by man and the great apes," *Advances in the Study of Behavior* 9 (1978), twenty sections of the paper deal with sexual dimorphism, the testes, the penis, seminal vesicles, and semen in gorillas, orangutans, chimpanzees, and humans. The female genitalia are scarcely mentioned and are covered in four summary statements at the end of each long section on male anatomy.

18. Useful reviews of this literature can be found in Symons (note 8). *The Hite Report* (note 8) provides concise summaries of the clinical studies but references to the anthropological literature in that book are not well balanced. Authoritative reviews of human sexuality from a cross-cultural perspective can be found in W. H. Davenport, "Sex in cross-cultural perspective," in *Human Sexuality in Four Perspectives,* ed. F. Beach (Baltimore: Johns Hopkins, 1977), and in *Human Sexual Behavior,* eds. D. S. Marshall and R. C. Suggs (New York: Basic Books, 1971).

19. See references to Barash, Beach, Pugh, Alexander and Noonan, and myself in Chapter 7, and references to Burton, Lancaster, Chevalier-Skolnikoff, Zumpe, Michael, and Goldfoot in notes 20, 21, 25, and 26 for this chapter. Elaine Morgan's ideas on female orgasms can be found in chapter five of *The Descent of Woman* (London: Souvenir Press, 1972).

20. Doris Zumpe and Richard Michael, "The clutching reaction and orgasms in the female rhesus monkey *(Macaca mulatta),*" *Journal of Endocrinology* 40:117–123 (1968). For competent reviews of the literature on female orgasm in primates, see also Symons' book (note 8) and Jane Lancaster, "Sex and gender in evolutionary perspective," in *Human Sexuality: A Comparative and Developmental Perspective,* ed. Herant Katchadourian (Berkeley: University of California Press, 1979).

21. Frances Burton, "Sexual climax in female *Macaca mulatta,*" *Proceedings of the Third International Congress of Primatology* 3:180–191 (1971).

22. Masters and Johnson (note 16), p. 52.

23. Hite (note 8), pp. 190–192.

24. G. Saayman, "The menstrual cycle and sexual behaviour in a troop of free-ranging chacma baboons (*Papio ursinus*) under free-ranging conditions," *Folia Primatologica* 12:81–100 (1970). W. J. Hamilton and P. C. Arrowood, "Copulatory vocalizations of chacma baboons (*Papio ursinus*), gibbons (*Hylobates hoolock*), and humans," *Science* 200:1405–1409 (1978).

25. Richard Michael, M. I. Wilson, and D. Zumpe, "The bisexual behavior of female rhesus monkeys," in *Sex Differences in Behavior,* eds. R. C. Friedman, R. M. Richart, and R. L. Vande Wiele (New York: Wiley, 1974); Suzanne Chevalier-Skolnikoff, "Male–female, female–female, and male–male sexual behavior in the stumptail monkey, with special attention to the female orgasm," *Archives of Sexual Behavior* 3:95–116 (1974).

26. Chevalier-Skolnikoff (note 25), p. 109; J. K. Hampton, S. J. Hampton, and B. T. Landwehr, "Observations on a successful breeding colony of the marmoset *Oedipomidas oedipus*," *Folia Primatologica* 4:265–287 (1966). Philip Hershkovitz, *Living New World Primates (Platyrrhini)*, vol. 1 (Chicago: University of Chicago Press, 1977), p. 769. It is worth noting that some women experiencing orgasm pant and make a face similar to the round-mouthed expression given by macaques. The incidence of this response is not known, and I know of no research which describes in detail facial expressions during human orgasm nor attempts to examine the origins of such responses.

27. Akers and Conaway, reviewing the literature on female homosexual mounting among rhesus, captive pigtailed, Japanese, and stumptail macaques and among chimpanzees, squirrel monkeys, and wild vervet monkeys, stress the effects of both hormonal factors (typically, the mounted female is in the ovulatory phase of her cycle) and close affiliative relationships between the females involved. J. H. Akers and Clinton Conaway, "Female homosexual behavior in *Macaca mulatta*," *Archives of Sexual Behavior* 8(1):63–80 (1979). Similar observations have been made for caged pygmy chimpanzees and wild gorillas.

28. Chevalier-Skolnikoff indicates that female responses during heterosexual copulations are less intense and external manifestations of "orgasm" less obvious than they are in homosexual mounts. Some—for example Symons (note 8)—believe this finding contradicts the claim that stumptail macaque females experience orgasms during both homosexual and heterosexual interactions, since the clear pattern of orgasmic response observed during homosexual mounting are not seen during heterosexual copulations. It is worth noting, however, that Chevalier-Skolnikoff addresses this very problem by pointing out—through extrapolation from the human data—that intensity of female orgasm and its manifestations may vary with

circumstances. "Masters and Johnson [1966]," she writes, "have found that the intensity of the human female orgasm is variable, mild orgasm being hardly distinguishable behaviorally or physiologically, while intense ones involve dramatic behavioral and physiological changes. They have also found that the more intense clitoral stimulation tends to produce a more intense orgasmic response than vaginal coitus. In view of this variability in the human female orgasm, it is conceivable that in stumptail females less intense and less easily identifiable orgasm than those observed during homosexual interactions might occur during heterosexual copulation." Chevalier-Skolnikoff (note 25), p. 113.

29. D. A. Goldfoot, H. Westerborg-van Loon, W. Groeneveld, and A. Koos Slob, "Behavioral and physiological evidence of sexual climax in the female stump-tailed macaque (*Macaca arctoides*)," *Science* 208:1477–1479 (1980). See also Eric Phoebus, "Coital heart rate in the female rhesus monkey (*Macaca mulatta*)," Ph.D. thesis presented to University of California, Irvine (1977).

30. Linda Wolfe, "Behavioral patterns of estrous females of the Arashiyama West Troop of Japanese macaques (*Macaca fuscata*)," *Primates* 20(4):525–534 (1979).

31. The most detailed account concerns wild orangutans in Sumatra. "Several wild orang utans, especially youngsters, were observed to stimulate their own genitals, either manually or by means of inanimate objects. Female orang utans might masturbate by rubbing their fingers, their foot or an object along their clitoris . . . or they might insert their hallux or objects into their vagina. One adolescent female . . . was observed to suck and wet the finger she used during her masturbation," H. D. Rijksen, *A Field Study on Sumatran Orang utans (Pongo pygmaeus Abelii Lesson 1827): Ecology, Behaviour and Conservation* (Wageningen: H. Veenman and B. V. Zonnen, 1978), pp. 262–263. Caroline Tutin reports similar behavior for a young female chimpanzee: "Gremlin showed a fascination with her own genitals during her fourth year, manipulating them directly with her hand and also rubbing objects, such as stones and leaves against them," C. E G. Tutin, "Sexual behaviour and mating patterns in a community of wild chimpanzees (*Pan troglodytes schweinfurthii*)," Ph.D. thesis presented to University of Edinburgh (1975), p. 139. Comparable accounts for captive animals include several for the monogamous tamarins (*Saquinus oedipus*) in which hand-tamed females masturbate either with their tails or against soft surfaces in the environment until stopping in a "trancelike" state, or until "satisfied." See reports by Hampton et al. and Hershkovitz, cited in Hershkovitz (note 26), p. 769.

32. Current hypotheses to explain the existence of female orgasms include: (1) Orgasms are therapeutic either because, as Sherfey suggested, they relieve "venous congestion," or because, as

Margaret Hamilton has suggested (unpublished manuscript), uterine contractions during intercourse somehow prepare the female for childbirth. (2) Orgasm enhances the probability of insemination either by transporting the sperm, as suggested by Fox et al., or by stimulating the male to ejaculate. Among some primates (many macaques, for example) males must mount, intromit the penis, and thrust many times before ejaculation is possible. This might be viewed as a reason for females to continue to be receptive and to solicit males for prolonged periods. This argument is by no means persuasive, however, since if it were the case that mechanical pecularities of the male constituted the chief selection pressure for female sexual traits, any male capable of quick ejaculation would outbreed these slower suitors. Hence, the answer in this case must be the other way around: males in these species take longer to ejaculate because prolonged stimulation somehow increases the likelihood of ovulation by the female or else other responses by the female conducive to eventual fertilization by this male.

As for the sperm-transport theory, the relevant data are contradictory. In an experimental study of the effects of orgasm on sperm transport, Masters and Johnson made radiographic check plates of six subjects (all of whom were orgasmic during the experimental session) to determine the direction of transport of a radio-opaque facsimile of seminal fluid which had been placed over the cervix of each subject. In none of the six individuals was there evidence of "the slightest sucking effect" of the semen facsimile, leading Masters and Johnson to conclude that orgasmic uterine contractions are "expulsive, not sucking or ingestive in character." Similar results had been obtained in a study by Bardwick and Behrman in which balloons had been extruded during uterine contractions. However, the additional finding that ejection via uterine contractions was most characteristic of subjects who were "anxious" raises the possibility that there may be differences among women and that women may respond differentially under different circumstances, and in particular, highlights the general difficulties of interpreting findings on this particular point on the basis of clinical experiments. Whatever the results, it is obvious from the existence of viable populations in cultures where females rarely or never experience orgasms during intercourse that orgasms are scarcely essential in order for fertilization to take place. The same—and several other—objections can be raised for Desmond Morris's idea (3) that orgasms keep females flat on their backs after sex, lest the seminal fluid flow out. See C. A. Fox, H. S. Wolff, and J. A. Baker, "Measurement of intra-vaginal and intra-uterine pressures during human coitus with radiotelemetry," *Journal of Reproductive Fertility* 22:56–76 (1970). Masters and Johnson (note 16), pp. 122–123. J. M. Bardwick and S. J. Behrman, "Investigation into the effects of anxiety, sexual arousal, and menstrual cycle phase on uterine contractions," *Psychosomatic Medicine* 29(5):470–482 (1967).

33. Owen Lovejoy, "The Origin of Man," *Science* 211:341–350 (1981). On dimorphism in polygynous Old World primates, see T. Clutton-Brock, P. Harvey, and B. Rudder, "Sexual dimorphism, socionomic sex ratio and body weight in primates," *Nature* 269:797–800 (1970); R. D. Alexander, J. L. Hoogland, R. D. Howard, K. Noonan, and P. W. Sherman, "Sexual dimorphism and breeding systems in pinnipeds, ungulates, primates and humans," in *Evolutionary Biology and Human Social Behavior,* eds. N. A. Chagnon and W. Irons (North Scituate, Mass., Duxbury Press, 1979). M. W. Wolpoff provides evidence for sexual dimorphism among fossil hominids in "Sexual dimorphism in the australopithecines," in *Paleoanthropology: Morphology and Paleoecology,* ed. R. H. Tuttle (The Hague: Mouton, 1975). See also D. C. Johanson and T. D. White, "A systematic assessment of early African hominids," *Science* 203:321–329 (1979); J. G. Fleagle, R. F. Kay, and E. L. Simons, "Sexual dimorphism in early anthropoids," *Nature* 287:328–330 (1980).

34. According to Kinsey et al. (note 4), pp. 375–376, 14 percent of females in their sample regularly responded with multiple orgasms. Some females had two, three, or as many as a dozen or more orgasms in a situation in which their husbands ejaculated only once. See also Masters and Johnson (note 16), p. 65.

35. In a recent survey of 106,000 readers of *Cosmopolitan* magazine, 23 percent of the respondents answered that they had had sex with more than one partner at a time; for most of these, sex with multiple partners meant two males. Of these respondents, 9 percent replied that they occasionally experimented with multiple partners; 14 percent replied that they had done so once. Linda Wolfe (note 4), p. 263. In the ethnographic literature, one of the very few known cases where women actually cite desire for multiple or sequential orgasms as a factor influencing the number of sexual partners is recorded for the aboriginal people of Western Arnhem Land, Australia. Women said to be dissatisfied by single male ejaculations may seek coitus with several different (marital and extramarital) partners within the same 12-hour period. R. M. Berndt and C. Berndt, "Sexual behavior in Western Arhem Land," *Viking Fund Publications in Anthropology,* vol. 16 (1951), p. 57.

36. The belief that female sexuality and reproductive capacity is somehow "dangerous" is widespread in various cultures throughout the world, and is often manifested in the special sanctions surrounding menstruating women. See for example Paula Weideger, *Menstruation and Menopause: The Physiology and Psychology, the Myth*

and Reality (New York: Alfred A. Knopf, 1976); and J. Delaney, M. J. Lupton, and E. Toth, *The Curse: A Cultural History of Menstruation* (New York: E. P. Dutton, 1977).

37. Alice Schlegel, *Male Dominance and Female Autonomy* (New Haven: Human Relations Area Files Press, 1972), p. 88. See also Gwen Broude, "Extramarital sex norms in cross-cultural perspective," *Behavior Science Research* 15(3):181–218 (1980).

38. Friedrich Engels, *The Origin of the Family, Private Property, and the State* (New York: International Publishers, 1942; English translation of the 1884 German original), pp. 54–55. At about the same time (1865) the anthropologist John McLennan was pursuing similar ideas: "The blood-ties through females being obvious and indisputable the idea of blood relationship, as soon as it was formed, must have begun to develop . . . into a system embracing them . . . But blood-ties through fathers could not find a place in a system of kinship, unless circumstances usually allowed of some degree of certainty as to who the father of a child was." *Primitive Marriage: An Inquiry into the Form of Capture in Marriage Ceremonies,* ed. Peter Riviere (Chicago: University of Chicago Press, 1970, reprint).

39. Nancy Marval, "The case for feminist celibacy," 1971 pamphlet, cited in Hite (note 8), p. 152. Sherfey (note 5, 1966).

40. See for example Jeffrey A. Kurland, "Paternity, mother's brother, and human sociality," in *Evolutionary Biology and Human Social Behavior* (note 33), p. 176.

41. Robert van Gulik, *Sexual Life in Ancient China: A Preliminary Survey of Chinese Sex and Society from ca. 1500 B.C. till 1644 A.D.* (Leiden: E. J. Brill, 1974), p. 189.

42. Mildred Dickemann, "Female infanticide and the reproductive strategies of stratified human societies: a preliminary model," in *Evolutionary Biology and Human Social Behavior* (note 33), pp. 321–367; "The ecology of mating systems in hypergynous dowry societies," *Social Science Information* 18(2): 163–195; "Paternal confidence and dowry competition: a biocultural analysis of purdah," in *Natural Selection and Social Behavior,* eds. R. D. Alexander and D. Tinkle (Concord, Mass.: Chiron Press, 1981).

43. Sarah C. Blaffer, *The Black-man of Zinacantan: A Central American Legend* (Austin: University of Texas Press, 1972), pp. 34–35; 118–119.

44. From Dickemann, "Paternal confidence" (note 42).

45. Mildred Dickeman, "Demographic consequences of infanticide in man," *Annual Review of Ecology and Systematics* 6:107–137 (1975). (Note that M. Dickeman and M. Dickemann are the same person; she changed the spelling of her name in 1978.)

46. Agatharchides' observations are cited by Diodorus and

Strabo. These and other early observations are summarized in Carl Gosta Widstrand, "Female infibulation," *Studia Ethnographica Upsaliensia* 20(varia 1):95–122 (1964).

47. Fran P. Hosken, "Female circumcision and fertility in Africa," *Women and Health* 1(6):1–11 (1976). It should be noted that counted in this large estimate are operations varying greatly in severity. In some cultures the clitoris is clipped rather than excised, and there is some dispute over the extent to which clitoridectomy reduces orgasmic capacity.

48. A. Abu-el-Futuh Shandall, "Circumcision and infibulation of females: a general consideration of the problem and a clinical study of the complications in Sudanese women," *Sudan Medical Journal* 5(4):178–212 (1967). Further consideration of the medical consequences of circumcision are discussed in J. A. Verzin, "Sequelae of female circumcision," *Tropical Doctor* 5:163–169 (1975); R. Cook, *Damage to Physical Health from Pharonic Circumcision (Infibulation) of Females: a Review of the Medical Literature,* report from regional advisor of World Health Organization to Division of Family Health (Geneva: WHO, 1976).

49. For example, see John Hartung, "On natural selection and the inheritance of wealth," *Current Anthropology* 17(4):612–613 (1976), or Soheir Morsy, "Sex differences and folk illness in an Egyptian village," in *Women in the Muslim World,* eds. Lois Beck and Nikki Keddie (Cambridge: Harvard University Press, 1978), pp. 610–611.

50. Carol L. Cronin, "Dominance relations and females," in *Dominance Relations: An Ethological View of Human Conflict and Social Interaction,* eds. D. R. Omark, F. F. Strayer, and D. G. Freedman (New York: Garland Press, 1980), esp. pp. 299, 302–303, and 317. See also notes 40 and 41 in Chapter 6.

51. Cronin (note 50), pp. 317–318.

52. I am indebted to Mildred Dickemann, who pointed this out to me. Dickemann lays out the theoretical groundwork for this view in the series of three papers listed in note 42.

53. Naomi Quinn, "Anthropological studies on women's status," *Annual Review of Anthropology* 6:181–225 (1977).

54. Anthropologist Louise Lamphere writes that "ethnographic reports show that many kinds of domestic groups are ridden with conflict and competition between women. Accounts of jealousy among co-wives, of the dominance of mother-in-law over daughter-in-law, and of quarrels between sisters-in-law provide some of the most common examples." She reviews this material in her article "Strategies, cooperation, and conflict among women in domestic groups," in *Woman, Culture and Society,* eds. M. Z. Rosaldo and L. Lamphere (Stanford: Stanford University Press, 1974). Among the more remarkable findings in this area are those of Robert LeVine,

who analyzed jealousy and hostility between co-wives in polygynous societies. LeVine found that polygynous societies where co-wives lived close to one another, in the same house or in adjacent households sharing a compound, were more liable to exhibit high levels of hostility between women as manifested by more frequent accusations of witchcraft. In a cross-cultural survey to test his hypothesis, he found a strong association between polygyny and sorcery, and this effect appeared to be magnified by having women live close by rather than in separate residences. "Witchcraft and co-wife proximity in southwestern Kenya," Ethnology 1 (1):39–45 (1962).

55. Lorna J. Marshall, *The !Kung of Nyae Nyae* (Cambridge: Harvard University Press, 1976), p. 280.

Index